The
Illustrated Dictionary
of Building Terms

The Illustrated Dictionary of Building Terms

Tom Philbin

McGraw-Hill

New York San Francisco Washington, D.C. Auckland Bogotá
Caracas Lisbon London Madrid Mexico City Milan
Montreal New Delhi San Juan Singapore
Sydney Tokyo Toronto

McGraw-Hill

*A Division of The **McGraw·Hill** Companies*

pbk   1 2 3 4 5 6 7 8 9 DOC/DOC 9 0 0 9 8 7 6

Library of Congress Cataloging-in-Publication Data
Philbin, Tom
   The illustrated dictionary of building terms / by Tom Philbin.
      p.   cm.

   ISBN 0-07-049729-X
   1. Building — Dictionaries.
I. Title.
TH9.P45 1996
690'.03—DC20   96-35305
                                                          CIP

McGraw-Hill books are available at special quantity discounts to use as premiums and sales promotions, or for use in corporate training programs. For more information, please write to the Director of Special Sales, McGraw-Hill, 11 West 19th Street, New York, NY 10011. Or contact your local bookstore.

Acquisitions editor: April Nolan
Editorial team: Executive Editor: Lori Flaherty
                Supervising Editor: David McCandless
Production team: DTP supervisor: Pat Caruso
                DTP operators: Tanya Howden, Kim Sheran, John Lovell
                DTP computer artist supervisor: Tess Raynor
                DTP computer artists: Nora Ananos, Charles Burkhour,
                   Steven Jay Gellert, Charles Nappa
                Proofreading: Cindi Bell
Design team: Designer: Jaclyn J. Boone
                Associate Designer: Katherine Lukaszewicz   DICT6

Introduction

This book is engineered to help today's builders and contractors. Hopefully, it will do that in a variety of ways. For one thing, it can be used purely as a building terms dictionary. It covers a wide variety of product, material, and technique terms. Most of these could be encountered in everyday work, but there are a few offbeat and arcane ones to add a little flavor. You never know when someone will wonder if they can have work done on their "entablature," for example.

But this is much more than a dictionary. It is an encyclopedia, and as such, it fleshes out important entries with information that includes everything from guidance on what is available to what is good and bad to how the material is installed. In more than a few cases, it includes a little history of a product or a technique, which is not only interesting to know but brings something to life.

Plywood is one such example. It was invented around the turn of the century by a gentleman in Portland, Oregon named Gustav A. Carlson. Carlson bonded some boards together with foul-smelling animal glue and displayed the product at the World's Fair. Despite this significant drawback, the product survived, until the real heroes of plywood—lab technicians—started to bind those veneers with phenolic resins, which were not only strong, but waterproof—and did not smell. Then along came World War II, and the need for plywood in the construction of PT boats—and plywood never looked back.

There is also, wherever possible, etymology, how these terms came to be. Where, for example, did the Phillips head screws come from? If you guessed from someone named Phillips, you would be right. Henry F. Phillips, to be exact, who lived in Portland, Oregon and invented the screw in 1934. (Portland did pretty well!)

The book covers many standard terms, but it also includes the argot of the profession because on the job words are crafted and tooled to serve one's purpose. In here, you can find everything from "mud jobs" to "polish coats" to "alligatoring."

And there is also pure slang, X-rated—but still commonly used in the industry.

Occasionally, my golden words fail, but there are plenty of illustrations to close the know-how gap.

In sum, I have tried to write a book that would teach me, but that I would also enjoy reading.

Tom Philbin

ABS Acronitrile butadiene styrene. A type of plastic used to make plumbing pipe. *See also* WATER PIPE.

absorption field A field engineered to receive SEPTIC TANK effluent. Also called a *leaching* or *seeping field*, an absorption field consists of a series of shallow trenches—parallel, round, or whatever the land allows—18″ to 24″ wide, in which are placed DRAIN TILES, pieces of perforated pipe. If the permeability of the soil, established by something called a PERCOLATION TEST is exceptional, then the pipes can be laid directly on it. In most cases the trenches must be lined with a 6″ layer of gravel to help the effluent absorb properly into the soil. *See also* CESSPOOL. *See illustration on following page.*

abutment Supporting column for an arch.

AC Alternating CURRENT. Describes electrical current or electrons that reverse direction because the generators that produce them reverse positive and negative poles constantly—60 times a second, or 1500 miles per second.

AC is the most common electrical current used in homes. As it travels along WIRES to a home or building, there is a minuscule drop in its force, but it is "stepped up" by transformers so it can continue its journey and stepped down by other transformers so it can be safely used. For example, it may start its journey from a power plant at up to 765,000 VOLTS, and be stepped all the way down to 120 volts for use.

It is this ability to be stepped up and down that makes it superior to DC, or direct current. *See also* ELECTRICAL SYSTEM.

Edison versus Tesla

For all his genius, Thomas Edison took the wrong road when it came to electrical current. He was a champion of direct current; another scientist, Nikolai Tesla, championed alternating current, which proved to be far more practical.

The competition proved intense, with Edison engaging in a number of experiments to prove DC was better, including electrocuting stray dogs.

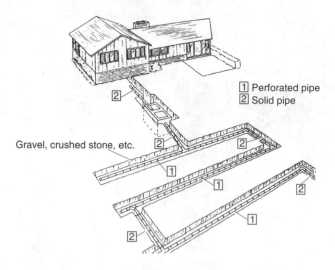

[1] Perforated pipe
[2] Solid pipe

Gravel, crushed stone, etc.

Absorption field

In the end, though, when Edison opened a massive generating station in Niagara Falls, New York, he had to license technology from Tesla.

Early electricity

The gas used to light lamps in American homes in the 1800s wasn't the clean-burning natural gas of today; it was smelly, sooty, and dangerous. When electricity came into American homes, the people were more than ready to embrace it.

Early systems were imperfect, to say the least. Poorly insulated wires, supported by wooden knobs or cleats, ran on wall faces. The wood was ultimately replaced by porcelain or glass when the current-carrying capacity of wet wood was discovered.

At one point, the wires were installed inside walls in a system known as "knob and tube." Knobs were attached to studs and the wires were strung along them; wherever the wire had to pass through a framing member, a porcelain tube kept it from contacting the wood.

Early INCANDESCENT lights were often mounted on gas lamps that were frequently required to act as backup lighting. When electric lights became more reliable, wire was run through the CONDUITS used for gas. Eventually, separate conduits were developed for wires, including Greenfield, which eventually transmogrified into BX where the conduit came with the wires already in place.

accelerator A substance that shortens the setting times of CONCRETE and MORTAR. Accelerators, or *accelerants,* are added in cold weather when there is a danger of the concrete or mortar freezing before it can CURE properly. Accelerators are considered ADMIXTURES.

acoustical tiles Tiles that reduce sound within a room. Many people think that acoustical materials lower or stop sound transmission between rooms. While this is true to some degree, to achieve soundproofing between rooms requires installation of other materials, such as heavy DRYWALL and INSULATION. Tiles, as well as other acoustical material, lower sound within a room by absorbing it.

Acoustical material will also not effectively prevent sound from entering a room. Part of the reason why acoustical materials are effective in muting sound is that they are lightweight, and noise generated within a room passes through them; noise coming the other way does the same thing.

Acoustic tiles have been used in America since at least the 1920s. *Scientific American* magazine reported in its September 1924 issue "Professor Saline invented an acoustical tile that is many times as absorbent as the usual masonry surfaces."

Tiles may be part of a suspended ceiling, or secured to FURRING strips, which are nailed to the ceiling. The tiles have TONGUE-AND-GROOVE edges which slip into each other. Staples are commonly used to secure them.

acrylic resin synthetic Resin used as a bonding agent or sealer in CONCRETE construction.

active solar heater A mechanical system for collecting solar energy. The water is heated in solar collectors, then stored in tanks for distribution within the building.

actual size The actual size of a piece of material, as opposed to its NOMINAL SIZE. BOARDS and timbers are one size when "rough," and smaller after milling or finishing. For example, a 2″ × 4″ nominal size is only 1½″ by 3½″ actual size. BRICK and BLOCK also have actual sizes different from their nominal size.

With many other materials, the nominal size is the same as the actual. For example, if someone speaks of a 4′ × 8′ sheet of PLYWOOD, that is exactly what it will be. In general, the actual and nominal size of most materials is the same.

addition Chemical added to cement during its manufacture either to help the manufacturing process or change its use characteristics. Additions are often incorrectly called ADDITIVES.

additives A cementitious substance added to PAINT to change it into sand or stucco paint. Additives are available in small boxes. Paint also comes with the additives already mixed in.

Also, chemicals added to paint to promote drying, deter insects, control viscosity, flow, and gloss levels.

adhesive Man-made product used to BOND a wide variety of materials.

Although they may do the same or similar jobs, adhesives differ from glue in one fundamental way: adhesive is made from synthetic materials while glue is made from organic material, such as horses' hooves.

Many adhesives have broad-base applications; one can use the same adhesive on a variety of jobs. But some are very specific. For example, adhesive used to adhere foam board must be of a type designed to use on the material or there is a very real danger of chemicals in the adhesive eating the foam. *See also* CONSTRUCTION ADHESIVE, CONTACT CEMENT, *and* THIN SET.

The same thing—in different packages

While many adhesives should be used only in particular situations, a number—although ostensibly different—are just the same product in different packages. For example, silicone caulk, although it is individually labelled for specific things such as concrete, metal, and wood, is the same product in each case.

The reason, of course, is increased sales: the customer buys four products for four different purposes, rather than one of the same product for all four purposes.

---

admixture Substance added to CONCRETE, STUCCO, or MORTAR for any of a variety of purposes: air entrainment (*see* AIR ENTRAINED CONCRETE), coloring, or to change drying time. Stucco, for example, is difficult to trowel unless a small amount of LIME is added to the mix. *See also* ACCELERATOR.

adobe BRICK made by drying in the sun rather than by oven firing, as is done with standard bricks.

Adobe brick, still used for building in certain areas of the country, may be made and used virtually anywhere as long as the bricks are made properly. To make the bricks, which are generally larger than standard brick, requires a certain kind of clay containing between 25% and 45% aluminum salts. The clay holds together a sand portion of the mix, and straw is added so bricks can shrink without damage. The final ingredient is emulsified ASPHALT, which is added for waterproofing and which was first developed by the Babylonians centuries ago. (This aspect of their craft was lost for a long time but was rediscovered about seventy years ago.)

Water is added and the mix is poured into molds and allowed to dry in the sun. In New Mexico and contiguous areas where the bricks are not exposed to severe weather, they are commonly stuccoed over and are not treated with emulsified asphalt. When completed, adobe brick must be up to code just as standard brick is.

after-sheathing window Any WINDOW that is installed after the SHEATHING is installed. Carpentry-wise, this is more difficult than installing the other type of window, the BEFORE SHEATHING one. The after sheathing type requires the carpenter to nail through the casing and sheathing into the JACK STUDS; the before sheathing type has a wood lip that one nails through directly to the studs.

aggregate A sand or stone component used in the making of CONCRETE. Aggregate is essentially a filler material in making concrete, which consists of PORTLAND CEMENT, water, and the aggregate. Aggregate generally makes up 75% of the bulk. Depending on the POUR—what is required strength-wise—aggregate can range in size from sand (fine aggregate) to pieces of stone (coarse) up to 2½″ wide. Aggregate does not add much strength to the concrete mix, but a poor aggregate—the wrong size stone and/or one of poor quality—can weaken the mix considerably.

air chamber A capped piece of pipe partially filled with air near a plumbing FIXTURE. Air chambers, also called air cushions, are designed to absorb the shock from abrupt water stoppage, a phenomenon known as "water hammer."

Air chambers cut down on noise, and can also help prevent damage to a PLUMBING system, parts of which can loosen if continuously shaken.

air content The volume of air in CONCRETE, CEMENT, or MORTAR. *See* AIR ENTRAINED CONCRETE *and* AIR POCKET.

air entrained concrete CONCRETE that has been altered chemically to disperse air bubbles throughout the mix. Astonishingly, a yard of air-entrained concrete may contain over three trillion bubbles. Such a mix is easier to work than standard concrete, stands up well in cold weather, and resists salts better than other concrete. It is particularly good for pouring in cold weather because the bubbles allow the concrete to expand and contract more readily. On the other hand, it is not as strong as regular concrete.

air pocket Dead air space in concrete, also known as a void. Such pockets are usually one millimeter; entrained pockets are smaller. *See also* PUDDLING.

alkali Chemical bases in CONCRETE and MORTAR, such as soluble salts.

Alkalis may be present in concrete and can lead to expansion and cracking when in contact with AGGREGATE. Alkalis can also interfere with paint application, so it is always advisable to wait 60 days or so before painting, when all the alkalis have emerged from the concrete and can be cleaned off, rather than applying the paint, only to have the alkalis drive it off when they migrate to the surface. *See also* EFFLORESCENCE.

alkyd Synthetic resin used in the making of PAINT and other coatings. It is solvent-thinned and used in both interior and exterior paint.

> ### A giant step forward
>
> The development of alkyds around World War II was a giant step forward in the making of paint. For hundreds of years beforehand, the heart of paint was LINSEED OIL; alkyds were a major improvement over this. Paints made with alkyds cover better, resist chalking and mildew better, dry faster, and withstand sunlight better.
>
> Alkyd- and oil-based paint have come to mean the same thing in some peoples' minds—both are thinned with mineral spirits or turpentine—but they are different. Oil-based usually refers to paint with a linseed oil base, while alkyd paint has an alkyd resin base. Linseed oil and alkyds are sometimes blended together in paint.

alligator 1. Dried PAINT that has cracked and taken on the segmented appearance of alligator skin. Alligatoring, sometimes known as crocodiling, is one of the banes of painting and usually occurs when a paint is applied to a glossy painted surface without first scuff-sanding it to promote adhesion, or when not enough time has been allowed between coats of paint or when a too-soft primer is used under a finish coat. **2.** ASPHALT top coating that cracks. Here, the cause is usually that a poor base has been laid.

American National Standards Institute A nonprofit group that tests building materials like the American Society of Testing Materials. (*See* ASTM *and* BUILDING CODE.)

amp *See* AMPERE.

ampacity *See* AMPERE.

ampere A basic unit of electricity, the amount of current one VOLT can send through one OHM. Ampere, commonly known as amp, is named after Andre-Marie Ampere, a French physicist. Amperage (how many amperes) is a key consideration in electrical power because it describes the amount of electricity flowing through the wire, the same as gallons per minute would describe the amount of water a pipe can carry. VOLTAGE is the push. If too much amperage is pushed through the wire, it can result in loss of power, or the wire's heating up and damaging insulation or causing a fire. *See also* ELECTRIC SERVICE.

anchor blocks Wood members secured to MASONRY structures to provide a nailing surface.

anchor bolt BOLT used to secure wood SILL plate to a MASONRY floor or FOUNDATION wall. These bolts, also known as "L bolts," are placed in the concrete while it is still wet. They are about eight inches long, and are set upright and about twelve feet apart with the hooked part down and at a point where the sill plates that have holes drilled in them can be slipped into place. Enough of the threaded portion of the bolt will be exposed to accept washers and nuts once a TERMITE SHIELD has been installed. *See illustration on following page.*

angle bead A metal or wood strip installed in the corner of a room to be plastered to protect the corner against damage and/or to serve as a guide to PLASTER placement.

angle bonds Metal TIES used to hold corners of MASONRY work together.

angle bracket Right-angle-shaped piece of metal designed to reinforce jointed boards.

angle iron A strip of structural iron bent to form a right angle and used to support MASONRY over openings.

approved Term used to indicate that a particular installation has been found to be in line with required regulations (the BUILDING CODE) by a governing

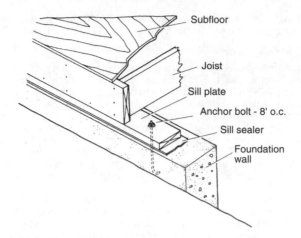

Anchor bolt

body. An installation is also deemed to be "up to code." For example: follow-ing the installation of a plumbing system, a plumbing inspector must declare it approved for use. *See also* BUILDING CODE.

apron 1. In carpentry, the piece of trim beneath a window STOOL. It helps support the window and gives a finished look. **2.** In paving, the area directly in front of a garage, and where the driveway intersects with the street or sidewalk. **3.** The front of a bathtub from the rim to the floor, also known as "skirting." **4.** Panel in back of a sink or lavatory.

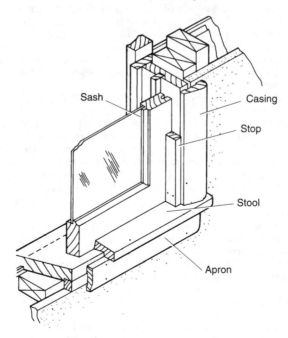

Apron

arcade 1. A vaulted place, open at one end of both sides; an arched opening or recess in a wall. **2.** A series of ARCHES either open or closed with MASONRY, supported by COLUMNS or PIERS.

arch Curved construction that spans an opening. Arches commonly consist of *voussoirs*, also known as ARCH STONES; or a pointed or curved member that is supported at the ends. Arch design varies greatly, from flat to elliptical.

arch brick 1. Wedge-shaped BRICK designed to be used in building an arch. **2.** Very hard, overburned brick.

architrave 1. In common use, collective name for the various parts (LINTEL, JAMB, and their MOLDINGS) that surround a door or window. **2.** In classical architecture, an architrave refers to the bottom section of an entablature, the portion of a building above a COLUMN.

arch stone Wedge-shaped stone, also called a *voussoir*, used in building an ARCH.

asbo Short for asbestos. Also known as 'bestos. Until the late seventies, when it was found to be dangerous, asbestos, which is waterproof and has high tolerance to heat, was used in a great many products. Perhaps surprisingly, it is still used in making some products, such as powdered floor levelers.

When is asbestos dangerous?

Asbestos has created a storm of controversy over the last decade or so because of its health hazards. In fact, asbestos is dangerous only in its unencapsulated form, when the fibers that comprise it float free in the air and can be breathed in, lodging in lungs and breathing passages. Hence, the best policy for encapsulated asbestos—wrapped tightly around pipes, for example—is to leave it alone.

asbestos runner Rope made with asbestos, wrapped around a pipe to facilitate the pouring of molten lead. Also called a "pouring rope."

ash dump In a fireplace, a grate that may be opened to allow fireplace ashes to fall into an ashpit below. An ash dump provides an easy way to store ashes until removal.

ashlar Squared stone used in building facings, FOUNDATIONS, and sidewalks. *See also* ASHLAR MASONRY.

ashlar masonry Masonry that has been assembled in different sizes and in random fashion to imitate ASHLAR.

asphalt A bituminous substance found in many parts of the world; a smooth, hard, brittle, black or brownish-black resinous mineral consisting of a mixture of various hydrocarbons. Called mineral pitch, "Jew's Pitch," and in the Old Testament, "slime." Asphalt can also be manufactured from products obtained in petroleum refining.

Asphalt is waterproof, and as such, it is an ingredient in many building products such as ASPHALT CEMENT, BUILDING PAPER, driveway SEALER, waterproofer, and ASPHALT ROOFING. *See also* BITUMEN.

asphalt cement Asphalt paving. Asphalt comes in varying grades of quality, and those in the trade say that the blacker the asphalt, the better its quality—there should not be any streaks of white or brown. Because it must be installed hot, the warm months of the year are best.

The most common job for the home is the installation of a driveway. To do this, the contractor first evaluates the soil for drainage. If the soil drains well, all the contractor has to do to is install a minimum of 4″ of well compacted asphalt. If the soil is clay, the contractor will dig it out and lay in a well compacted bed of gravel 4″ to 7″ thick. Over this will go the 4″ of asphalt.

Number one in complaints

Asphalt driveways are close to being number one—or are number one—for customer complaints in a number of states.

Cracking is usually the problem. This is normally caused by poor preparation of the bed in which the gravel is laid. Water gets under the base, freezes, and moves the asphalt. When asphalt moves, it cracks.

A too-thin layer of asphalt is also a problem. Contractors may install a 4″ base of gravel, but only 2″ of asphalt; 4″ is much better.

Another problem is puddling, which can be caused by a dip in the asphalt (known as a BIRDBATH) or by failing to "crown" it properly (sloping it up near the middle so water runs toward the sides).

asphalt roofing shingles Shingles whose major ingredient is ASPHALT.

Asphalt roofing shingles, and the companion material, FIBERGLASS ROOFING SHINGLES, collectively are on more than 80% of the roofs in America. They are simple to install, look good, wear well and are relatively inexpensive.

The shingle used may be 12″ to 36″ long and have cutouts along the bottom. These are called *tabs* or *flaps* and are the portion of the shingles that will be exposed to the weather. Most shingles come with dabs of adhesive on them to help keep them down in case of high winds.

All roof shingles also come rated for their ability to resist fire, the ratings being A, B, and C with A the highest. They come in various weights, according to how many pounds are in 100 square feet (a SQUARE) of area. For example, 215 pound shingles would weigh 215 pounds per square, and 235 pound shingles would weigh 235 pounds per square. Today, manufacturers only release such information to architects and contractors. Shingles are sold by the length of the warranty.

Asphalt shingles come with various warranties, up to 25 years, but many installers feel this is a gimmick. Customers will grow tired of the color or the roof will be so worn that they will opt to change it well before the 25 year warranty runs out.

As with a number of other products, it is important for the installer to follow the manufacturer's instructions exactly when installing asphalt shingles. A deviation from the prescribed procedure can lead to voiding the warranty.

Installation of asphalt shingles starts with an examination of the roof, and an awareness of how many "roofs" the local building code will allow. (In some localities you are allowed to install up to three layers of shingles before a TEAROFF down to the DECK—the roof's substrate—becomes necessary.) Sometimes the roof can be simply repaired and new shingles installed. For example, many of the shingle tabs may be cupped. These "scales" can simply be broken off and a new roof laid on them. Gaps can be filled with HORSEFEATHERS. If roofers see that a roof is in poor shape, they might also check to see if the deck is damaged, but this is rare. One roofer estimates that only one in 1000 roofs requires a new deck.

If the new roofing goes over existing material, the old roof will serve as a base. If not, and a tearoff has been done, the first step is the installation of BUILDING PAPER. The roofer lays this on the roof, overlapping the strips and securing them with large-headed nails or staples. Building paper provides some measure of weathertightness. The roofer also replaces or repairs FLASHING as needed.

Once the building paper is in, the shingles can be installed. The starter strip, an upside-down course of shingles secured along the edge of the roof, is installed. Another course is installed on top of the first course, and then the roofer proceeds towards the roof peak, overlapping the courses as he goes until he reaches it. He then works his way from the bottom of the roof on the other side until he reaches the peak. The peak, or ridge, is capped with short shingle sections.

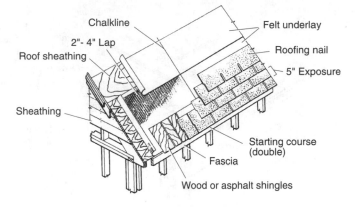

Chalkline
2"- 4" Lap
Roof sheathing
Felt underlay
Roofing nail
5" Exposure
Sheathing
Starting course
(double)
Fascia
Wood or asphalt shingles

Asphalt roofing shingles

The procedure above is for a straight GABLE-style roof. Many roofs have complicated angles and VALLEYS, and installation is somewhat complicated, but the essential methodology remains the same. *See also* DECK *and* ROOFING.

ASTM American Society for Testing Materials. Non-profit organization made up of a variety of professionals who decide the level of quality that building materials must have to be used on particular jobs.

Once a standard has been proposed, the qualified individuals in the ASTM either approve it or suggest changes. Once all the changes have been worked out, membership votes on the standard. If it is adopted, it is published and manufacturers are expected to comply with the standard when making their products. If manufacturers do not comply, then their products will not be used for jobs that require products made according to ASTM standards. Each product receives an ASTM number. *See also* AMERICAN NATIONAL STANDARDS INSTITUTE *and* BUILDING CODE.

astral A small molding, semicircular in section, sometimes plain, sometimes carved with leaves or cut into beads, placed around the top or bottom of columns, and used to separate the various parts of the ARCHITRAVE in ornamental entablature.

attic ventilator A screened opening provided to ventilate an attic space. Attic ventilators are located in the SOFFIT area as inlet ventilators and along the gable end or along the roof RIDGE as outlet ventilators. Fans may also be installed to aid attic ventilators. *See illustration on following page.*

> **The importance of ventilation**
>
> Ventilation bleeds off excess water vapor in the air that otherwise might condense (*see* CONDENSATION) and attack various areas of the house.
>
> If a home is covered with wood shingles, ventilation is more or less automatic: vapor passes through the shingles harmlessly into the atmosphere. ASPHALT or fiberglass ROOFING, on the other hand, creates a virtually airtight top to a house. Unless the warmed air has some means of escape, it can condense on rafters and the like, or melt snow and create ice dams, which can lead to further water damage.

In the summer, most of the heat will collect in an attic. Ventilators can bleed this off and help keep the heat from migrating into lower floors.

avonite A solid-surface material with a textured granite or gemstone look. Like other solid-surface materials, it is "color through"—the color is

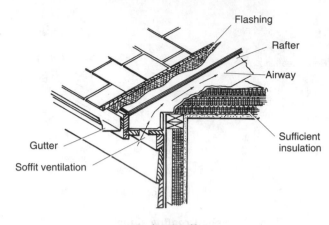

Attic ventilator

the same throughout the material. Avonite looks like stone but can be worked and polished with woodworking tools. It is a popular material for use in kitchens and baths.

Avonite is described by its manufacturer as a homogeneous blend of polyester alloys and fillers. *See also* CORIAN, FORMICA, *and* FOUNTAINHEAD.

awning window Window with a single panel of glass hinged at the top. They can be obtained "stacked"—several windows running vertically. The window has a bar or crank operator, which allows it to be operated without removing the screen or storm sash. The hinges should be the kind that allow space for one's arm between the SASH and frame so the window can be cleaned.

Awning windows can be left open during a light rain. They are also good over a kitchen sink because they can be opened without straining.

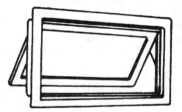

Awning window

backbrush To apply paint or stain with roller or spray, and then work it into the surface with a brush.

Some materials, such as exterior stains, work best if they are driven into the pores of the wood, a job for which a brush is the best tool. Many stain failures can be traced to inadequate or nonexistent backbrushing. *See also* STAIN.

backer rod foam Strips used to partially fill gaps before CAULK is applied; also called backup.

Caulk alone is not used to fill deep gaps because it is not designed to be applied so thickly. Foam rope does this job quite well, and a relatively thin topping of caulk completes the job. Mortite, a semi-solid caulking strip, may also be used, as can oakum, a heavy, chemically-saturated rope.

backfill 1. Earth, stone, and other material used to fill space around the foundation of a newly built house. **2.** The process of replacing excavated earth around and against a foundation wall.

It is important when backfilling that the earth be compacted and sloped properly. Compacted earth, as opposed to backfill just dumped in loosely, will retain its shape and ensure that water will flow away from the foundation wall.

Backfill is a blend of earth and stone and debris such as broken brick and block. It should never contain materials that can deteriorate easily, nor wood scraps, which can attract ruinous insects such as carpenter ants. *See also* EXCAVATION. *See illustration on following page.*

back filling 1. Filling over the extrados of an arch. **2.** Rough masonry behind a FACING or between two facings. **3.** Brickwork between structural members. Also known as brick nogging. **4.** *See* BACKFILL.

backing BLOCK wall used to serve as a base for a BRICK wall. The wall is built with an occasional HEADER brick tying the brick wall to the backing.

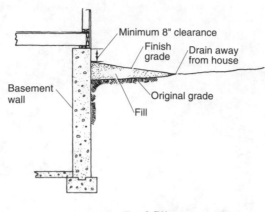

Backfill

back-mop Apply ASPHALT to back side of roofing felt during installation of a built-up ROOF.

backnailing Nailing piles of built-up roof (*see* ROOF) to substrate to supplement BACK MOPPING.

backplastering Application of a coat of PLASTER to the opposite side of a plaster wall after the plaster or MORTAR on the other side has hardened.

backpriming Applying paint or stain to the backs of siding. Many painters swear by backpriming, saying that it ensures the stability of the wood by keeping moisture from invading through the back side of the sections. Today one can also purchase some materials, such as cedar shakes (*see* SIDING), PRIMED both on the front and the back.

backset *See* DOOR LOCK.

backsplash The edging at the back of a kitchen COUNTERTOP to prevent sink water from splashing on or spilling down the adjacent wall.

backstamp American Plywood Association approval stamp on plywood panels. All unsanded and touch-sanded panels and panels with A or B faces on one side only carry the APA trademark on the panel back. Also known as an edgemark.

backup 1. Portion of a MASONRY wall behind the FACING. **2.** Overflowing of a plumbing FIXTURE due to a stoppage. **3.** Material installed before a sealant to help fill space and support sealant. *See* BACKER ROD FOAM.

badigeon A mixture of PLASTER and stone applied to a plaster wall to give it the appearance of stone.

bag Short for a bag of cement, which weighs 94 pounds.

ballast 1. Stone or gravel used as a base for concrete. **2.** Device in a FLUO-RESCENT lamp that provides starting and operating current. In essence, a ballast is a small transformer, regulating current flow to the pins or sockets on the tube.

ballcock The mechanism in a toilet tank that controls the flow of water. The standard ballcock is a tubular metal device threaded at one end to mount in the toilet tank. The other end has a plunger-like valve that actually regulates water flow. The valve is controlled by the rise and fall of the FLOAT BALL.

There are literally hundreds of makes of ballcocks around, making it a monumental task to obtain the right one when something goes wrong. This is why plumbers will ordinarily replace the entire mechanism; ballcocks are universally available in sizes that will fit most tanks.

The diaphragm ballcock is a plastic tube with a plastic canister on top that houses a diaphragm mechanism. This ballcock operates by WATER

PRESSURE, and is quieter than a standard ballcock. It can be installed in most toilets. It comes in two models, one of which has an anti-siphon feature to ensure that waste water will not be siphoned into the water supply system.

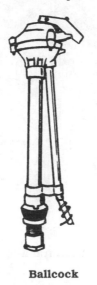

Ballcock

balloon framing A method of framing a building, the essential characteristic of which is STUDS that extend from basement to ROOF. The building generally follows other methods. A foundation is made, and then 2 × 6s, or SILLS, are bolted to the top of it after laying in a MORTAR BED and making sure the sills are level. Then POSTS, usually made with 2 × 4s that are spiked together, are secured at the corners, with the post prepared so that there is recess for accommodating interior wall materials. JOISTS are secured to the sill next. These are made of "TWO BY," 6" to 14" deep. They are laid across the sill and then the house-high studs are raised into position and PLUMBED, and spiked to the sill and adjacent joists using the same spacing. _See illustration on following page._

Ribbands, or ribbons, are let in and secured to the studs at the second floor level on opposite sides of the house; these boards, 1' × 6' or 1" × 8", are used to support the joists, which are laid across the top of and secured to studs. A pair of ribbands is also secured to the studs at the ceiling/attic height of the second floor joists. PLATES are then secured to the tops of the studs, each plate consisting of 2 × 4s that have been spiked to the studs and then to each other. The plates are cut to fit where they contact the corner joists.

As construction goes, balloon-frame construction is not as strong as PLATFORM or braced-frame construction, so the use of SHEATHING is critical to give it solidity and wind resistance. When high winds rack the house, the loads travel down the sheathing into the sill, which is much better able to withstand it. (_See_ WIND LOAD.)

BRACING is normally not used with balloon-frame construction when sheathing is employed, but it can be where risks are high—in hurricane country, for example. Firestops are essential in balloon-frame construction. These are merely 2 × 4s that are installed horizontally between studs at certain points. In a two-story home there should be two: one at the ribband at the sill, and the other at the second floor ribband. Without these firestops, spaces between studs could act as flues, and a fire would spread much more rapidly. _See illustration for firestops on following page._

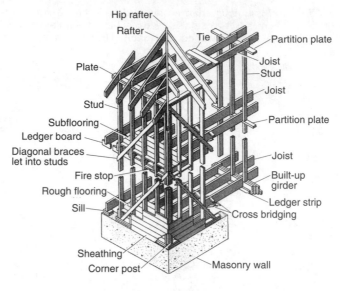

Hip rafter
Rafter
Tie
Partition plate
Joist
Stud
Joist
Plate
Partition plate
Stud
Subflooring
Ledger board
Diagonal braces
let into studs
Joist
Fire stop
Built-up
girder
Rough flooring
Sill
Ledger strip
Cross bridging
Sheathing
Corner post
Masonry wall

Balloon framing

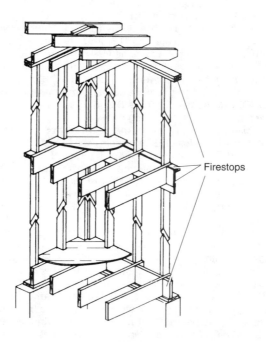

Firestops

Firestops

A revolution in framing

Balloon framing—the name refers to its light, airy nature—revolutionized building in America when it emerged in the 1830s, along with the manufacturing of NAILS by machine. Until balloon framing, builders used laborious TIMBER framing, which involved using relatively exotic joinery and large wood members that could range in size from 4″ × 4″ to 9″ × 15″. With balloon-framing, buildings could be erected at much greater speed and with far less cost (balloon framing members ranged from 2″ × 4″ to 2″ × 12″). Said *Architectural Review* magazine in 1945: "The great cities could not have arisen as quickly as they did if it were not for the invention of the balloon frame, which substituted a simple construction of nails and plates for the old craft or mortised and tenoned joints in wooden house construction."

A plentiful supply of NAILS was also crucial. For a long time, the only nails available to the carpenter were hand-wrought ones. To make a nail, a blacksmith had to heat a long thin metal rod called a nail rod, hammer one end to a point, cut the rod to make it nail-sized, and pound the unpointed end to a head. Making nails was so labor-intensive, the cost of building a house could be half of the total materials cost. Hand-wrought nails were so difficult to make and valuable that the nails were often salvaged when old structures were taken down.

Starting around 1800, nails were made by machine. These nails, called cut nails, were cut from sheets of metal: two sides were tapered and two straight; the head was flat. Unlike with the hand-wrought ones, carpenters did not have to worry about losing a few cut nails.

Today, nails are made from huge coils of wire (hence their name "wire cut"). One end is pinched off to a point, and the other end is flattened.

Nail size is still designated by the penny weight, the way it has been since the 1400s.

baluster Short pillar or column, slender on the top and bulging on the bottom to an elliptical or pear shape. The term comes from *baluster*, an Italian word meaning "blossom of the wild pomegranate," whose shape it resembles. *See also* BALUSTRADE *and* BANISTER. *See illustration on following page.*

balustrade A row of BALUSTERS surmounted by a rail or COPING, forming an ornamental PARAPET or barrier along the edge or a terrace, balcony, etc. *See also* BANISTER.

banister Slender upright post or rail, guarding the side of a staircase and supporting the handrail. The word is a corruption of BALUSTER.

banjo vanity A vanity cut in the shape of a banjo.

bar In masonry, short for rebar. Rebars are metal rods of various lengths and diameters with a deformed or raised surface. Rebars are available for reinforcing any type of masonry construction. On concrete BLOCK construction, they are installed vertically in the CORES of the block, which are then filled with concrete. With the use of rebars, builders can often use concrete block construction in the place of pure concrete, saving money without sacrificing strength.

bargeboard The decorative slanting boards on the GABLE end of a ROOF. Whenever there is a roof overhang, bargeboards are used to cover the exposed wends of framing members. Bargeboards vary in fanciness. In homes built in the late nineteenth century, they are very fancy.

The word barge comes from a medieval Latin word meaning gallows. Back then, bargeboards were a key pair of timbers used in the construction of the gallows. Bargeboards are also known as gableboards and vergeboards.

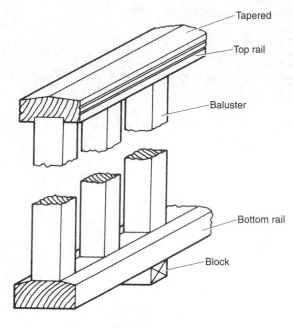

Baluster

baseboard Plain or fancy boards that run around the bottom of an inside wall and conceal the gap between floor and wall. *See* WALL.

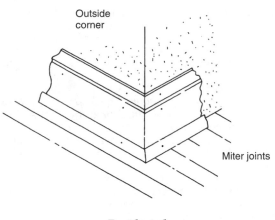

Baseboard

base cabinets In a kitchen, the cabinets that rest on the floor.

Base cabinets are installed first in kitchen installation or remodeling jobs. Making them level ensures that the COUNTERTOP will be level, so as they are installed, each is leveled by using shims (usually pieces of Number 2 undercourse shingles). If, when measured, the collective cabinets are more than ¾″ out of level then the carpenter will likely remove them and trim the bottoms, and reshim as needed.

Once the cabinets are level, a line is drawn along the wall following the top of the cabinets. The cabinets are removed and a nailer (usually 1¼″ × 2″) is screwed to the wall and the base cabinets are replaced, releveled and screwed to the nailer.

Finally, the cabinets are screwed together at the STILES. The next step would be to install the countertop. *See also* KITCHEN CABINET.

base course In MASONRY, the bottom-most course.

basement vent Glazed or screened openings in the top of the FOUNDATION wall. Basements and attics need ventilation more than other areas, and these vents help provide it for the basement. *See also* ATTIC VENTILATOR.

base molding Molding used to trim the upper edge of a BASEBOARD, where it meets the wall.

base shoe Molding strip used to cover gap between BASEBOARD-floor joint. This piece of molding is particularly useful if the floor is uneven.

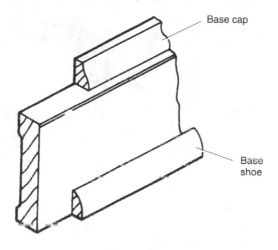

Base cap

Base shoe

Base shoe

bat Piece of BRICK with one end intact, the other broken off. Bats, also called brickbats, are used in wall-building where only partial bricks are needed.

bath, receptor Small, low TUB designed to be used as a shower base or tub for a child.

bath, recessed A TUB enclosed on three sides by walls.

Bathing not always popular

The Greeks and particularly the Romans elevated taking a bath to the level of high sensory experience. One could go to the local bathhouse and get not only a bath, but a steaming, a powdering, and an annointing as well.

Bathing fell on hard times in the the Dark Ages. Abstaining came to be regarded as suitable mortification for the flesh of the unholy person; the use of hot water, specifically, was condemned as self-indulgent. (Some noblemen and medieval kings did bathe occasionally. In the 13th century, for example, it is reported that King John of England bathed at least three times a year. Queen Elizabeth is reported to have taken a bath at least once a month "whether she needed it," one of her ministers said.) The most important anti-bathing concept during those days was simple: bathing was dangerous to your health.

Even in Colonial times, bathing was not all that popular. Indeed, in a few fledgling states—Ohio, Virginia, and Pennsylvania—legislation banned or restricted bathing. Ben Franklin was an exception to the rule: he bathed regularly. His peers, still wedded to European non-bathing traditions, criticized him, giving him the vicious label of "The Father of American Bathing."

Bathing only came into popularity in America in the 19th century.

batten 1. In general, a narrow strip of wood. **2.** Narrow strips of wood or plywood used to hide joints between panels. **3.** Strip of wood used in making a door. Here, battens are securd horizontally across vertical members in the making of a batten door. **4.** In roofing, strips of wood used as a base or DECK for slate, clay tiles, or wood shingles. **5.** In PLASTERING, used as LATH. *See also* BOARD *and* SIDING.

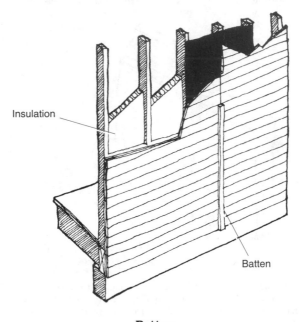

Insulation

Batten

Batten

batter A wall purposely constructed to slope on the outside and be straight and true inside. The term may come from warfare, when hundreds of years ago battering rams were used to knock defensive walls down.

batter board One of a pair of horizontal boards nailed to posts set at corners of an EXCAVATION, used to indicate the level wanted, or used as fastening posts for stretched strings to indicate outlines of FOUNDATION walls. *See illustration on following page.*

battleships Metal strips used in mounting electrical switches and receptacles in OLD WORK that do not have some other hardware to keep the box from falling out of the wall, such as built-in clamps. Battleships are technically known as Madison clips, but in profile look like a battleship, hence the name. *See illustration on following page.*

batts Precut lengths (usually four or eight feet) of INSULATION consisting of a filler of mineral wool or fiberglass and facings, one of which is a VAPOR BARRIER, and flanges for nailing to the framing members such as STUDS or ceiling BEAMS.

Batts are usually filled with fiberglass in glass fiber form and resemble cotton candy (Indeed, some children have eaten it). MINERAL WOOL, popular in the 1950s and 1960s, was supplanted by fiberglass-filled batts because mineral wool soaks up moisture, reducing or negating its insulative value.

Insulation is the sane quality—good or bad. The defining factor is the R-VALUE of the insulation. The higher the number, the better the material is. Fiberglass generally has an R-value of 3.7 per inch of thickness; and mineral wool, a 3.

The term likely comes from "batt," which is defined in the *Oxford English Dictionary* as "a felted mass of fur, or of hair and wool in hat-making." *See also* BLANKETS.

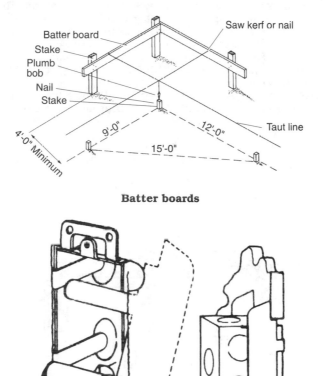

Batter boards

Battleships

bay window Window that, when viewed from the top, has one straight and two angled sides. A bay window increases floor space because, in effect, it is like adding a small extension onto a house. New flooring as well as ROOFING material will be required. *See also* BOW WINDOW.

bead 1. General term for applying a strip of material, such as a line of CAULK.
  2. A raised length of woodwork.

beam A general term used to describe a large horizontal supporting piece—wood or metal—in a building. Beams are used where solid strength is required, such as supporting a load over an opening or floors.

Before BALLOON FRAMING came along, most houses in America were POST-AND-BEAM constructed, meaning that the heavy posts comprised the vertical members while a variety of beams supplied horizontal support. When beams were first used, they could be huge; an entire tree could be used for one beam.

beam ceiling Ceiling construction where horizontal members, or beams, are exposed. Beams on this type of ceiling may be functional or purely decorative, but in either case are exposed to view.

beam fill Masonry used to fill spaces in some buildings to provide FIRESTOPS.

bearer In general, any supporting member of a house. Bearer is sometimes used to describe specific supports. For example, the piece of wood that helps support a WINDER, a turning step in a staircase.

bearing The part of a BEAM or TRUSS that is in direct contact with its support.

bearing plate Steel plate placed on a GIRDER, COLUMN, or TRUSS beam to distribute the weight of the load.

bearing wall A wall helps support the weight of a building in addition to its own weight. If a house has a RIDGE line, a bearing wall would normally run at a right angle to it. It can be difficult to identify which are bearing walls and which are not.

If a bearing wall were removed from a house, the house would not likely collapse, but it would sag measurably and cause significant damage. *See also* PARTITION WALL *and* WALL.

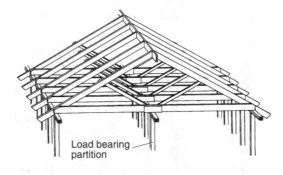

Load bearing partition

Bearing wall

bed 1. A layer of mortar into which brick or stone is set. Also known as a bed joint. **2.** A fiberglass or other rot-resistant material applied to the top of a foundation before SILL boards are installed. Installed, sill boards may not be perfectly flush with the top of a foundation wall, and this could provide an entry point for water. Bedding helps ensure that gaps are sealed. **3.** To set a pane of glass in place on GLAZING COMPOUND. Such a bed can be useful because of the normal expansion and contraction that wood goes through. As the window frame expands, it pushes the glass against a relatively flexible material such as glazing compound rather than the wood or metal, and avoids cracking the glass.

bed joint *See* BED.

bed molding Strip of wood used to conceal the joint between an exterior SOFFIT and wall. Bed molding serves the same purpose as a BASE MOLDING, except it is at the top of a wall rather than the bottom.

bed stone A large stone used as a foundation for a GIRDER.

before sheathing window Trade lingo for a WINDOW that is installed before the SHEATHING is installed. This type of window is more difficult to install than its AFTER SHEATHING counterpart. It has an extra strip of wood on the side that you nail through directly to the studs to secure.

belt course Narrow horizontal BRICK course, slightly projected from the face of the rest of the masonry, such as a window SILL. Also called a sill or string course.

'bestos Lingo for asbestos. *See also* ASBO.

bevel cut **1.** Board or other material cut at an angle. **2.** Plywood panels cut at an angle to make smooth mating joints.

bevel siding Commonly known as clapboard. In cross section, clapboard is angled, hence the name. *See* SIDING.

bib Short for bibcock, a FAUCET with a nozzle that is bent downward.

bidet Bath fixture for cleansing the perineal area of the body following excretion. Although it can be found in America, the bidet has not caught on as a FIXTURE like it has in Europe.

How the bidet got its start

Bidet is a French word meaning "small horse" or "pony." It gained popularity among the soldiers of Napoleon's armies who created a wooden structure they could straddle and wash themselves on after many hours on horseback, providing relief and preventing rashes.

The bidet was misunderstood and maligned for awhile. In the early 1900s, the Ritz-Carlton Hotel in New York City installed a few but were forced to remove them by enraged moralists of the time. In 1950, the Hotel Moliere tried to explain the use of the fixture to foreigners with a sign above the fixture which read "Foot Bath."

---

birdbath A depression in an ASPHALT driveway filled with water.

bird's mouth cut A cut made in a RAFTER end that resembles the open mouth of a bird. The bird's mouth cut (also called "notch" or simply "bird's mouth") is made so the rafter joints are snug with the top wall PLATE. It looks like an open bird's mouth, hence the name. Some construction does not require bird's mouth cuts but most does.

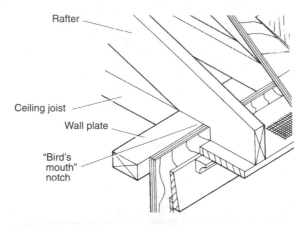

Rafter

Ceiling joist

Wall plate

"Bird's mouth" notch

Rafter with bird's mouth cut

bitumen The generic term for an amorphous, semi-solid mixture of complex hydrocarbons derived from any organic source. ASPHALT and COAL TAR are the two bitumens in the construction industry, being vital to roofing, coatings, and cements.

black iron Iron with no finish on it. It is actually gray-black in color.

blankets Long lengths (40 to 100 feet) of INSULATION filled with fiberglass or mineral (rock) wool of some type. Also known as "rolls," blankets are used when lengths of insulation over eight feet are needed. The area is measured and the blanket cut as needed, laid in place, and secured to framing members by stapling flanges to them.

The blankets come in various widths to fit between framing members from 16" to 24" apart. They also come in various thicknesses; the thicker the material, the greater the R-VALUE. Blankets come with a VAPOR BARRIER that is directed inside the house. If the blankets come unfaced, polyethelene sheeting is stapled to the studs once the insulation is in place.

As with BATTS, it is important that the flanges be snugly stapled to the framing members, particularly in colder climates. Gaps can lead to warm air from the inside passing through to the exterior materials, where it can contact cold air and condense. This water, unchecked, can do a lot of damage. *See also* CONDENSATION.

bleed/bleed through 1. When PAINT or another coating does not adequately cover, and the finish below can be seen. If a green wall were covered with white and a green cast showed through, it could be said that the green was bleeding through. **2.** Inadequately primed wall patches that bleed through finish paint. **3.** Patches primed with wrong primer. One might use a flat wall paint, repair an area with plaster, then coat the patch with shellac. The shellac would bleed through the paint. **4.** In MASONRY, the exuding of water from concrete after the masonry is in place.

blind nailing Nailing in such a way that nailheads are not visible in the finished work. Blind nailing is normally done at the tongues of TONGUE-AND-GROOVE boards, such as STRIP FLOORING and under ROOFING materials.

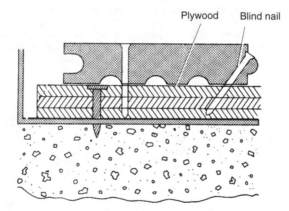

Plywood   Blind nail

Blind nail

block A building unit made of concrete whose NOMINAL SIZE is $8'' \times 8'' \times 16''$, but whose ACTUAL SIZE is $7\frac{5}{8}'' \times 7\frac{5}{8}'' \times 15\frac{5}{8}''$. The $\frac{3}{8}''$ is to allow for a MORTAR joint that is $\frac{3}{8}''$.

Today block is popular for building everything from a foundation wall to a complete building. Blocks weigh between forty and fifty pounds each, the weight variance mainly due to the different weights of the aggregate used. Block may be solid—which means they are totally solid or have small air cells in them—or contain voids, or "cores" (either two or three), to reduce the weight of the block.

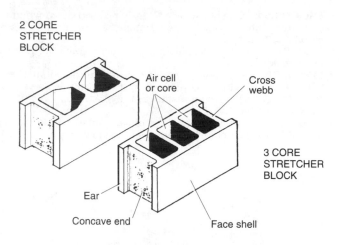

Typical block

Such blocks are characterized as a heavyweight type. For residential use, "lightweight" blocks are popular, though at 30 to 40 pounds each, they are not truly light. These are made with an aggregate such as expanded shale or clay, pumice, perlite or the like rather than stones. Still strong enough for the job, they are easier to handle.

Although most blocks are very strong and can be used for most purposes, whether they are suitable for a particular application should be checked with a supplier before purchase. Blocks also come graded according to where they can be used. Type N blocks are designed for use above or below grade, whether subject to water penetration or not. Type S blocks are designed for use inside, where they are protected from the weather. The basic difference between the two types of block is the aggregate used.

Blocks come in a variety of shapes to facilitate building without having to cut them. Stretcher blocks, for example, are used to build the main section of a wall, while bullnose blocks are used where rounded corners are needed. Jamb blocks are used to accommodate doors.

There are also "screen" and "grill" block, which are decorative and have large openings. Erected, they work like a fence.

Block may also come glazed, or with the faces worked or scored in some way. For example:

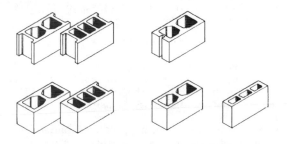

Various others shapes of block

- **Slump block.** The blocks are slightly pressed down when they are made to create the squat appearance of adobe. From a distance, they cannot be told apart.
- **Split.** The face shell is split, exposing aggregate and making the block look like stone.
- **Fluted.** The block bears a series of scoop marks along the face.
- **Scored.** Block with a score mark in it that imitates a vertical MORTAR JOINT. From a distance, the block looks as if it is built with smaller block—8″ × 8″ × 8″.
- **Striated.** The block has a series of comb-like striations along its face. It looks particularly good when shadow hits it.

An architectural block

Cured—or not?

Concrete shrinks as it dries, and it is very important for the builder to know whether the block he uses is or isn't cured. Shrinkage in uncured block can be from ½″ to ¾″ per 100 feet, something that can easily create cracks in a finished wall.

Concrete blocks will also swell if they become wet, so during and after curing, it is important to keep them protected. Once installed, dry block that is properly cured will become wet, but this will not be significant in terms of structural integrity.

Cinder block—just the name survives

In 1917, Francis J. Straub received a patent for a block that used an aggregate made of cinders, which in huge supply because so much coal was being burned around that time. The cinders were fireproof and light, the perfect aggregate for a concrete block.

Cinder block sold extremely well for many years, but as people moved away from coal heat, the supply of cinders diminished. Other aggregates were used in place of them to produce a lightweight block.

Today, even though there are not many cinders around, and almost no block is produced with them as an aggregate, concrete block is still popularly known as cinder block.

blocking 1. Short boards installed between joists to strengthen a floor. Without blocking or BRIDGING, a simpler form of blocking, floor joists can twist and bend, leaving the floor too springy and unstable. Just how much blocking is installed depends on the type of floor and the distance spanned. **2.** In MASONRY, a means of building walls not built at the same time with OFFSETS whose vertical dimensions are not less than 8″. *See illustration on following page.*

bloom *See* EFFLORESCENCE.

blow A raised section; delamination in a plywood panel. Blows are caused during the manufacturing process, usually because of steam pressure building during the time the panel is manufactured. The steam itself may come from moisture within the veneer, excessive glue spread, or high press temperatures.

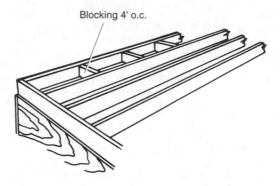

Blocking 4' o.c.

Blocks help strengthen joists

blown-in insulation Insulation that is "injected" into house walls or other areas.
Blown-in insulation is just that: blown in. SIDING sections are removed and holes are drilled in the SHEATHING to allow insulation to be pumped or blown into the wall cavities. Blown-in insulation is also commonly pumped into attic floors that are not accessible other ways, and between rafters.
Preparation for a blown-in insulation job involving attic floors or rafters means screening off—with fireproof material such as sheet metal—lights, chimneys, and other sources of heat so that the insulation is not in contact with heat. Even if the insulation has been treated with a fire-retardent chemical, it can burn.
There are a number of blown-in insulations, as follows:
- **Fiberglass.** This is glass fibers that, in installed form, are about the texture of cotton candy. Fiberglass has less insulation value than BATTS, BLANKETS, or BOARD insulation, but it is less costly to install per square foot. Fiberglass insulation is supposed to be vermin-proof, but mice have nested in it. (Perhaps they can't read.)
- **Cellulose.** This is macerated paper, usually newspaper, that has been treated with fire-retardent chemicals. Cellulose has a fairly high insulation value—higher than fiberglass—but with time, it can settle, allowing gaps in the protection. It can also soak up water. There is some concern, too, about the kinds of cellulose that contain aluminum sulphite. It is said that this compound can attack the FRAMING FASTENERS used on roof TRUSSES, and actually dissolve them and weaken the roof framing. The cellulose should be thoroughly checked out to see if this could create problems.
- **Urea-Formaldahyde.** This foam insulation is no longer in use; the formaldehyde in it is bad for some people's health. Before you purchase a home, the walls should be tested to see if there is any formaldehyde residue in them. EPA tests have indicated that the residue drops over time, but if it is a problem and has to come out, it is an expensive process. SIDING and SHEATHING are taken off to gain access, and then the foam is cut out.
- **Urea-Tripolymer.** This foam does not contain formaldehyde. It is applied like urea-formaldehyde. It tends to shrink, creating gaps in protection.

See also POURED-IN-PLACE INSULATION.

blueboard Water-resistant Sheetrock. Blueboard gets its name from the pale blue paper that covers it. It is commonly used as base for CERAMIC TILE. *See also* GREENBOARD.

blued A finish of limited protection, often given to nails to keep them from rusting while they are waiting to be sold.

bluestone blue-grey or green sandstone that splits easily into thin slabs. Bluestone is a variety of FLAGSTONE that is used for paving.

board 1. Softwood lumber (usually pine) with a NOMINAL, or named, thickness of less than 2″. Boards are normally available from 2″ to 12″ wide in two-inch increments (2″, 4″, 6″ etc.); one can also get sixteen-inch-wide boards, these made by gluing pieces together. *See also* ACTUAL SIZE, ONE BY.

board insulation Insulation in various rigid forms. Following is a lineup:

- **Polystyrene.** This is commonly called Styrofoam, which is the brand name of one company. For home and light building use, it is commonly available in 4 × 8 foot sheets and ½″ to 4″ thick. Light and easy to work with, polystyrene can be cut to fit between STUDS, or glued to MASONRY foundations and walls, or under siding. It can also be used under concrete floors. Polystyrene is waterproof; its most noticeable negative quality is that it can burn during a fire, causing toxic fumes.
  In recent years, polystyrene has come on as a material for building FORMS for concrete. See eps forms.
- **Rigid fiberglass.** This comes in various thicknesses and forms. Like polystyrene, it can be cut and glued to concrete walls. It is also made as panels for use in suspended ceilings, and comes foil-backed for use as house SHEATHING.
- **Polyurethane.** Although still available, polyurethane gained a bad reputation a number of years ago as a material that burns rapidly and produces toxic smoke. Although manufacturers say that they have added fire-retarding chemicals, the material, which comes in sheets and is applied with specialized machines, is banned in some communities unless it can be protected from fire.
- **Isocyanurate.** Although more expensive than any other insulation, isocyanurate has the greatest R-VALUE—double that of fiberglass. (One inch of rigid fiberglass has an R-factor of 3, while isocyanurate's is 7 or 8.) Isocyanurate comes 1 to 2 inches thick and in 2 × 6 and 4 × 8 sheets. It can be glued to foundation walls and the like, but its best use may be to insulate areas where other insulations cannot be installed thickly enough to achieve the desired R-factor.

boasted work Stone with lines chipped from it by a chisel.

bolster A short horizontal wood or steel beam on top of a column to support and decrease the SPAN of beams or girders.

bond 1. The pattern in which BRICKS are installed in a wall to ensure that vertical joints are not directly above one another to prevent a potentially weak wall. The arrangement is also important in terms of aesthetics—it gives a wall its distinctive look.

The different patterns are brought about by the way STRETCHER and HEADER bricks are arranged. (*See* BRICK.) A wide variety of bonds exist, as follows:

- **Common.** A variation of the running in brick with a header every fifth, sixth, or seventh COURSE. The header courses have joints in alignment.
- **English.** Alternate courses or stretchers. Headers are installed so the stretcher joints are centered under the header bricks.
- **Dutch.** Stretcher joints center on stretcher joints two courses above and below. Also known as the English Cross Bond.
- **Flemish.** The headers in one course are centered above and below the stretchers in the other course.
- **Running.** Uses stretcher bricks with the joints throughout occurring at the middle of bricks above and below a particular brick. Also known as a stretcher bond.
- **Stack.** This bond has joint lines aligned vertically. It is mainly used for decorative value because the aligned joints do not make for a strong structure. Roman style BRICK is normally used.

2. The joining of MASONRY materials with mortar. **3.** Materials joined with adhesive.

Running bond   1/3 Running bond   6" Course headers
Common bond   6" Course flemish headers
Common bond

Dutch corner   English corner   English corner   Dutch corner   English corner   Dutch corner

Flemish bond   English cross or Dutch bond   Stack bond   English bond

Brick bond patterns

bond beam Horizontal reinforced CONCRETE MASONRY beam designed to strengthen a masonry wall and reduce the possibility of cracking. *See also* BEAM.

bond breaker Material used between other materials to prevent them from bonding. For example, some concrete FORMS are coated with oil before CONCRETE is placed so it will not stick to the forms.

bondstone Stone used in a coursed rubble masonry wall. Rubble masonry walls are built of random-sized stones that are roughly squared, but for strength it is necessary to occasionally place stones that are big enough to go all the way through the wall. These are called bondstones. It is usually recommended that bondstones be placed at least once in every 6 to 10 square feet of masonry.

bonnet A cover used to guide and enclose the tail end of a VALVE spindle, or a cap over the end of a pipe.

boot 1. The connecting flange between a heating DUCT and the REGISTER. **2.** The flange around BASE of a pipe projecting through a roof.

boston ridge Method of applying shingles at the ridge or hips of a roof.

bow Distortion of a board or plywood panel so that it is bent rather than flat lengthwise.

bow window Large window that is shaped like a bow, looking at it from the top. The same things that can be said about a BAY WINDOW can be said about a bow type. *See also* WINDOW.

box 1. Pour paint back and forth between containers to ensure that it is mixed properly. **2.** Opening in a wall to accept an open shutter. **3.** Box that serves as a safety "container" for electrical connections.

When wires are stripped of insulation and wrapped around TERMINAL SCREWS, such as on a RECEPTACLE or SWITCH; or joined to a FIXTURE; or when they are joined for purposes of rerouting the direction of the CURRENT, they must be housed in a metal or plastic box for safety. No raw wires should be left exposed, not only because of the possibility of fire from electrical malfunctions, but also to protect anyone working on the wires in the future.

The same applies outside: any switch, receptacle or LAMP must have its electrical mechanism protected by a box, in this case a "weatherproof box," which may be plastic or metal. All localities allow for metal boxes, but some restrict plastic box uses. Boxes are available in various shapes and sizes. The NATIONAL ELECTRIC CODE, which many municipalities follow, places restrictions on how many wires may be in a box of a given size, because crowding of wires can be a safety hazard.

All boxes have screw-hole tappings in front to accept 6 × 32 machine screws (*see* SCREW) to secure switches and receptacles. All boxes also have knockouts, which are holes plugged with either metal or plastic discs.

Boxes also come with covers. There are plastic and metal covers with knockouts, slots for receptacles or switches, blank covers that screw on for a rarely used receptacle, or covers for "junction" boxes where wires terminate and connect to one another.

Following is a lineup of common boxes:

• **Four-inch square.** A four-inch square box is usually only 1½″ deep but sometimes found in a 2½″ deep box which is used when a wall is too shallow to accept a Gem box. While not as deep as a gem box, a four-inch square box has the same cubic capacity.

• **Gem.** A standard metal box for switches and receptacles. A gem box— no one seems exactly sure where the name came from—is two inches wide, three inches high, and 2½ inches deep, and can house a single receptacle or switch. If necessary, the sides of gem boxes can be removed and the boxes ganged—linked up—to form a large box. Gem boxes may be deeper than 2½ inches and have expandable brackets and plaster ears that are useful in mounting the box in OLD WORK.

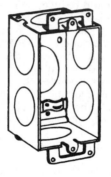

Gem box

• **Handy.** A surface-mounted box with its corners rounded off so that it will not scratch anyone who bumps into it. A handy box has screwholes in the back for it to be screwed to the wall.

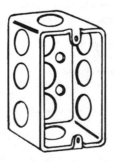

Handy box

- **Junction.** A four-inch-wide octagonal or round box for mounting a ceiling fixture in. A junction box—also known as a "ceiling box"—is either 1½ or 2⅛ inches deep with or without expandable arms. A junction box is for "new work."
- **New work.** This is designed to be used on new construction, where framing members are exposed and the box can be readily nailed in place. New work boxes usually have nailing brackets on them. The box is simply nailed to studs or other framing members as needed. The brackets are notched or otherwise constructed so that they tie in well with wall or ceiling material; they make the housed electrical device flush with the material.

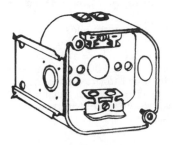

New work box

- **Old work.** This is designed to be used in an existing circuit where framing members are not exposed. Old work boxes require that holes be cut for them into the wall or ceiling material. Boxes are designed to grip the wall material so it can fit flush with the wall and not come loose. Some boxes have plaster ears, round metal projections that look like ears; some have clamps that can be spread out; and some have built-in screws or MADISON CLIPS.
- **Plastic one gang.** This looks like a gem box with expandable brackets. It is the only plastic box available for "old work," and it can be used to house either a receptacle or a switch.
- **Weatherproof.** This is for use outside a building. These boxes are made of cast aluminum or an alloy. Alloy weatherproof boxes have plates inside, with screw-hole tappings to screw switches or outlets to; cast aluminum weather boxes have built-in screw threads. Because regular steel screws react to aluminum and can corrode, stainless steel screws are used on aluminum boxes. (See GALVANIC ACTION.)

Weatherproof boxes use screw-on covers. When the box is not used for a long time, the covers are solid. Other times, the boxes have switch toggles and flip covers for receptacles; or you can unscrew the entire cover to get at the receptacle. The covers are weatherproof, not only in terms of the material they're made of, but also in their gasketed edges, which provide a watertight seal. Sometimes covers have ½″ threads to accept fixtures. These come prewired, ready for hookup inside the box. *See illustration on following page.*

box beam Beam built of lumber and plywood which make a long, hollow box. A box beam will support more load across an opening than will individual members. *See also* BEAM, BEAM CEILING *and* BREASTSUMMER, COLLAR BEAM *and* TRUSS.

box column Hollow COLUMN usually made of wood.

boxing out Marking off a building for location of electrical SWITCHES and RECEPTACLES. When framing is complete, the electrician will use crayon or chalk to mark the framing members with locations of switches, etc. The locations are double-checked by the architect or house builder, and the

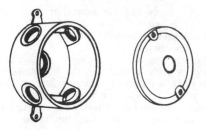

Weatherproof box with cover

outlet boxes that house the receptacles, switches, etc., are nailed to the framing members.

box sill A type of SILL plate. Here, there are headers which are nailed to, and circumscribe, joist ends. *See also* PLATFORM FRAMING.

brace Board diagonally installed to provide extra strength in a wood-framed structure. Braces are used in a variety of places in a building, but most commonly in a stud wall. Slots are cut in the studs near the corner and the braces are nailed in. *See also* BLOCKING.

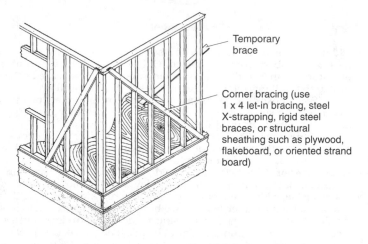

Temporary brace

Corner bracing (use 1 x 4 let-in bracing, steel X-strapping, rigid steel braces, or structural sheathing such as plywood, flakeboard, or oriented strand board)

Corner bracing

braced frame Bracing, usually diagonal, on buildings that use GIRTS that are mortised into solid posts.

bracket A support projecting from a wall to support a shelf or some ornamental feature, such as a cornice.

brad Finishing NAIL shorter than 1½". Tiny brads, are good for attaching delicate moldings; they are also used in making picture frames.

branch circuit In electricity, individual CIRCUITS that feed power to a house. Branch circuits exist because it would not be practical for a house to be served by a single circuit. If this were the case, then the entire house would lose power if one of the OUTLETS on the circuit lost it. Additionally, the size of the wire needed would be very heavy, making installation difficult. Under the general heading of branch circuits, a home will typically have three types: lighting, small appliance, and INDIVIDUAL APPLIANCE CIRCUIT.

brass Plumbing lingo for all FAUCETS and FITTINGS regardless of the material of which they are actually made. Brass is connected to rough plumbing.

brass-plated Steel that has been coated with a plating of brass. It is attractive and strong, and somewhat though not completely weatherproof. A clear lacquer spray does much to extend its life.

breaking joints Installing masonry or boards so vertical MORTAR JOINTS are not aligned. This results in a stronger structure than if the joints were in alignment. *See also* BOND.

breastsummer A large supporting horizontal beam in the face or breast of a wall. Breastsummer (also called bresoummer) is partly from the French word *sommier*, meaning BEAM. There is also some evidence in that the word derives from "sumter," a kind of large dog.

brick Building unit made with clay heated to extreme hardness. There are a variety of types and sizes, and the builders will commonly earmark particular kinds of brick for use in situations they are best suited for.

Brick can generally be broken down into four basic kinds: building brick, facing brick, firebrick, and paving brick.

Of the four, building brick is most commonly used—about 98% of all brick used is building, or common, brick. It comes in a great many styles and, like other brick, is available "solid" or "hollow." The hollow brick is lighter and generally allows for better mortar joints.

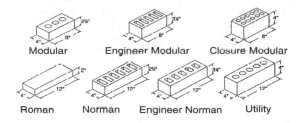

Modular   Engineer Modular   Closure Modular

Roman   Norman   Engineer Norman   Utility

Various brick

Building brick, like other brick, comes in various grades, related to the brick's ability to stand up to weather. There are three weathering grades. Grade SW is designed for use underground or where the climate is particularly cold and wet and frosty. Type MW are also geared to be used in a cold, wet climate, but only above ground. Type NW is used inside or outside, but wherever frost is not expected.

Facing type brick is for use wherever looks are important. This, in turn, is available in types SW and MW and in three facing grades. FBX is made to exacting standards with few flaws. While FBX brick is consistent in size and color type, FBX brick can have some variations among brick.

Firebrick is made with special clays and fired in ovens at high temperatures to make it particularly hard and fire-resistant. Such brick is commonly used inside a fireplace in homes.

Paving brick, or pavers, is specially prepared to be hard enough to walk on without causing undue wear.

Brick is also described in terms of usage or special characteristics, as follows:

- **Abrasive resistant.** Brick with low absorption, designed to be used in drainage situations.
- **Acid hesitant.** This type of brick can be used where acids are present. It is ordinarily used with acid-resistant mortar.
- **Angle.** Any brick shaped to an oblique angle to make it fit into a corner.

- **Arch. 1.** Wedge shape used in constructing arches. **2.** Very hard-burned brick.
- **Clinker.** Hard-burned brick whose shape is distorted or bloated due to complete vitrification.
- **Dry press.** Brick formed made from relatively dry clay.
- **Economy.** Brick with NOMINAL dimensions of 4″ × 4″ × 8″.
- **Floor.** Brick used as a flooring. Smooth and dense, it is highly resistant to abrasion.
- **Gauged. 1.** A tapered arch brick. **2.** Brick that has been manufactured to precise dimensions.
- **Hollow.** Brick with cores or voids. This makes for a lighter brick and, it is said, for better MORTAR joints.
- **Jumbo.** Brick larger than standard size. Some brick manufacturers use the term to describe brick specifically made by them.
- **Norman.** Brick with nominal dimensions of 4″ × 2⅔″ × 12″.
- **Roman.** Brick with a nominal dimension of 4″ × 2″ × 12″.
- **Salmon.** Describes underburned brick, usually orange-pinkish in color, slightly larger and more porous than hardburned brick.
- **Snap header.** Half a brick.
- **SCR.** SCR stands for Structural Clay Research and is the trademark of the Brick Institute of America. SCR brick has dimensions of 6″ × 2⅔″ × 12″.

Bricks are also described in terms of their placement in a wall, how they fit into or comprise the COURSES. Stretcher bricks are laid flat and parallel to the wall being built. Header bricks are laid across two walls and have an end facing front. Rowlock stretcher is laid parallel to the wall, but on edge. Rowlock is placed like a header, but on edge. A sailor position is when the brick is placed on end, but face-front. *See also* BOND, COURSE, *and* MORTAR JOINT.

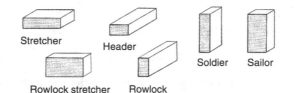

Stretcher   Header   Soldier   Sailor

Rowlock stretcher   Rowlock

Brick is known by different names depending on the position it is in a wall

For 10,000 years—and why

It is estimated that brick has been used as a building material for around 10,000 years. It is easy to see why. The clay used in making brick can be obtained almost anywhere, and making it is not very difficult, assuming one has the equipment. Moreover, the manufactured product is available in small units and can be made in a wide variety of shades and finishes, and it is strong enough to serve as the foundation of a building, yet handsome enough to add decorative accents.

Making brick

Although the process of making brick is done by machine, it is essentially done the way it has been for thousands of years. Suitable clay is mixed with water, placed in a mold, and dried. For years, drying was done by simply letting the brick air dry, a process that could take weeks; but today it is put in kilns and fired at temperatures up to 2400 degrees and can be completed in a few hours.

Depending on the process used, brick color may be varied, as may weatherproofness and other qualities. Vitrification is the key in determining brick strength, not the clay used. This refers to the clay particles being heated fusing together. Brick strength will depend on how this is done.

brick veneer Refers to WALLS that consist of BRICK tied to SHEATHING. Such brick has no structural strength, but depends on the sheathing (or studs) it is linked to. People commonly select brick veneer so their homes look like they are made of brick, without the cost.

Bricks and weather

Many people think that brick is impervious to water. In fact, this is not true. Bricks and MORTAR can absorb a large amount of water, but when the bricks are exposed to drier air, the water evaporates, and the brick returns to normal.

Curiously, older brick is often better at resisting water penetration than newer brick. Older bricks were more absorptive, but they also contained lime, which distributes the moisture better than the denser, stronger brick developed after World War II.

To get through a brick wall completely, it is estimated that a rainstorm would have to hammer a building for 24 hours with 50 mph winds. In areas with such conditions, a solution is to build a "drainage wall," also known as a CAVITY WALL.

Stronger when mortar is weaker

Curiously, a brick wall is stronger when the mortar is weaker. Contrary to fairly common belief, mortar is used simply to join the brick, not because it has great compressive strength.

Properly mixed and applied, mortar can and does develop a series of fine, invisible cracks when the wall is subject to stress, including when it settles. If the mortar were stronger than the brick, it would be the brick that cracks, which would weaken the wall. However, if the mortar develops a crack that follows the joint lines where brick actually separates, then that is an indication of poorly made or applied mortar, and the wall is weaker because of it.

bridging Small wood or metal members nailed in a diagonal position between the floor JOISTS at midspan to brace them. Bridging is cheaper than blocking because it requires less wood.

bright zinc A plating that protects items against tarnishing and gives them some degree of weather protection. Bright-zinc-finished items can be painted.

bronze-plated Steel that has been coated with a plating of bronze. This is somewhat weather-resistant, although it is usually used for attractiveness.

brown coat In two-coat PLASTER jobs, the first or base coat; and in three-coat jobs, the second coat. The brown coat—which gets its name from its brown color (it turns brown as it dries)—is normally applied directly to a base of MASONRY, GYPSUM LATH, or the other absorptive materials, but is used as the second coat if the base is wood or metal lath. To ensure stability in the latter, a SCRATCH COAT must first be applied.

The brown coat is the thickest of coats applied; it gives walls and ceilings most of their strength and solidity. It must be level and smooth, or it will be difficult to smooth the finish coat. Applying it is commonly referred to as "browning."

See also PUTTY COAT.

BTU British thermal unit. The heat energy required to raise the temperature of one pound of water one degree Fahrenheit.

buck FRAMING around an opening in a wall. A door buck, for example, describes the framing around a door.

buff To shine a surface.

building code Set of rules and regulations restricting how and with what materials a building may be constructed.

Building codes arose out of necessity in America. Before they came along, builders were free to throw together anything they wished, sometimes with fatal consequences. In New York City, early tenements were poorly constructed and poorly thought out. They not only contributed to substandard living, but calamitous fires which killed many people. Finally, in response to this, New York City established building codes.

Early building codes were of the type that specified what materials and methods builders could use. In a sense, they amounted to overkill. As time went by, more realistic codes came into being. These were simply based on whether the materials and methods performed adequately. You did not have to use cast iron if plastic did the job; you did not have to use BX if ROMEX cable was sufficient; you did not need PLASTER walls when Sheetrock was perfectly adequate.

The adoption of performance codes was greatly aided by two organizations: ASTM, the American Society for Testing and Materials; and ANSI, the AMERICAN NATIONAL STANDARDS INSTITUTE. Both organizations supply standards for products and all builders have to do is select products that comply. It gives them great leverage in building.

Today, building codes vary from locality to locality and are a blend of codes established by organizations, area tradition, and reality. Plumbing codes in the southwest say copper tubing may not be used because the HARD WATER there can lead to deposits that can clog the pipes. SOFT WATER can also be a problem for copper pipe. Environmental factors in building codes include such things as freezing in the winter, earthquakes, and hurricane stresses. There are a number of areas in the country where codes dictate that the house be able to withstand such horrific forces. *See* EARTHQUAKE LOADS.

Politics are another kind of reality. Contractors have their associations, which have been known to pressure legislative bodies to change this or that code to make things easier for the contractor. And, to some degree, things have changed.

> **Like a sore thumb**
>
> At one time during the early days of building codes, it was decreed that water towers in New York must be conical and made of wood staves. Today, remnants of that decree stand out "like a sore thumb," as one engineer put it, on the roofs of many Manhattan skyscrapers, which are otherwise made of masonry and are of modern design.

building lines Any of a variety of lines to mark off the parameters of a building. A building line can mark the excavation limits, or where the building face falls, or a building line can be an imaginary line extending from the corners of a building to mark a driveway. *See illustration on following page.*

building paper Material installed at various points in a house for waterproofing, and/or dampproofing, dust proofing, and windproofing. Building paper is commonly made of asphalt felts (or tarpaper) and is commonly installed under SIDING, between SUBFLOORS and finish floors of various kinds, behind PANELING and beneath ROOFING. *See also* TYVEK. *See illustration on following page.*

built-up roof Roof composed of three to five layers of ASPHALT felt laminated with COAL TAR, pitch, or asphalt with the top layer having crushed slag or GRAVEL. Depending on the method of application, such roofs are usually characterized as ten-, fifteen-, or twenty-year-old roofs. *See illustration on following page.*

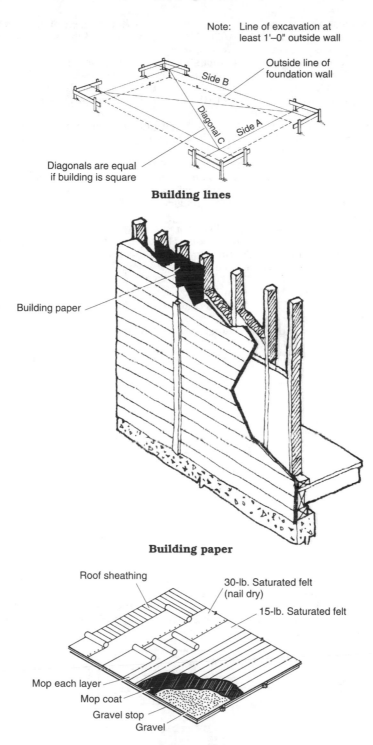

Note: Line of excavation at least 1'–0" outside wall

Outside line of foundation wall

Side B

Diagonal C

Side A

Diagonals are equal if building is square

Building lines

Building paper

Building paper

Roof sheathing

30-lb. Saturated felt (nail dry)

15-lb. Saturated felt

Mop each layer

Mop coat

Gravel stop

Gravel

Built-up roof

bulkhead 1. An inclined door that allows access to a basement. **2.** A roof structure over a stairway. **3.** In MASONRY, a partition that blocks new concrete being poured into one form from oozing into another.

bundle 1. Standard package of roof shingles. Usually contains 20 or so shingles. Three or four bundles are used per "square." (*See* ASPHALT ROOFING SHINGLES.) **2.** A unit or stack of plywood held together with metal bands.

burning the joint Tooling MORTAR after it has hardened. This process rubs metal off the tool, leaving dark marks on the mortar.

busbar Metal bar to which WIRES coming into the house are attached and from which CIRCUITS are routed. The term derives from the brand name Buss, a common brand of FUSES.

bushing In plumbing, a short, threaded piece of pipe used to link larger or smaller pipes. Bushings have male threads on one end and female threads on the other, and come in a variety of sizes. *See also* FITTINGS.

butt 1. The thicker end of a roofing shingle. **2.** A BUTT HINGE. **3.** Short ROOFING shingle.

butter 1. Apply JOINT COMPOUND, commonly referred to as "mud," to drywall panels, in a series of thin coats. The task is not a simple one; it takes a lot of experience to make the compound smooth. It is done by tradesmen who are known as tapers because they use joint tape to help cover up the seams. **2.** Apply MORTAR to BRICK or BLOCK prior to seeing it in position. The name comes from the consistency of the MORTAR used—it is hard and firm like butter. When a brick is buttered, the implication is that the brick is hard and dense and the joint thin.

butt hinge Hinge with two rectangular leaves with screw holes joined by a pin or rod. *See also* HINGE.

butt joint JOINT the mating surfaces of a material are perfectly square.

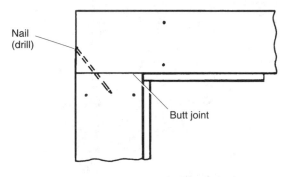

Nail
(drill)

Butt joint

Butt joint

BX A rigid, flexible metal CONDUIT containing two or three insulated wires, each wrapped in spiraled layers of tough paper.

Of the two types of electrical cable commonly used—BX and ROMEX—Romex is considered safer and more secure. Many electrical CODES permit only it to be used.

BX cable

How BX got its name

The term BX arose around the turn of the century. At that time there was a company run by two men, Gus Johnson and Harry Greenfield. Their primary product, GREENFIELD, looked just like BX but was hollow. It worked the way it works now: electricians snaked it through walls and then pulled wires through it.

One day while manufacturing the product, a mishap occurred that fed cord through it, and someone wondered: what if we were to run wire through it, instead of cord, and sell it complete?

A batch was made and given to an electrician to test. He came back raving about it, how easy it made the job of running wires—they were in place—and asked for as much as the company could supply.

He also wanted to know the product's name. One was invented on the spot. "BX," one of the partners said. "B because it's our B-product, and X because it's experimental."

The technical name for BX is armored CABLE.

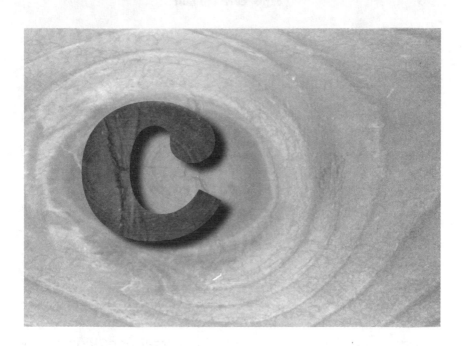

cabinet head Decorative molding piece that runs horizontally across the top of a door frame or ARCHITRAVE. Not many of today's homes feature cabinet heads, but in older buildings they were quite well featured, and made of carved and tooled woods.

cabinets *See* FACE-FRAMED CABINETS, EUROPEAN CABINETS, *and* KITCHEN CABINETS.

cable In electricity, consists of two or more CONDUCTORS or wires encased in a sheath of some sort. All wire in the interior of cables is the same, in a sense. It can carry just so much CURRENT. It is the exterior sheathing that determines exactly where the cable can be used. *See* WIRE *for specific types.*

cadmium-plated Refers to a steel item that has been coated with cadmium to make it rust-resistant. Many different hardware items are cadmium-plated, which is more water-resistant than zinc.

candlepower The intensity of light coming from a source in a particular direction. This is not a particularly meaningful way to measure light because it only covers one direction and not all directions the light travels. *See also* FOOTCANDLE; FOOTLAMBERT, *and* LUMEN.

> **Measuring candlepower**
>
> Today there is a variety of sophisticated instruments to measure a light's candlepower. When candles were made of whale oil the method was to measure not the light but the candle, which consumed wax at the rate of 120 grains per hour.

cant To place something at an angle.

cantilever A horizontal structural component that projects beyond its support, such as a second-story floor that projects out from the wall of the first floor.

cant strip Triangular piece of lumber used at the junction of a flat DECK and a wall to prevent cracking of the ROOFING applied over it. *See illustration on following page.*

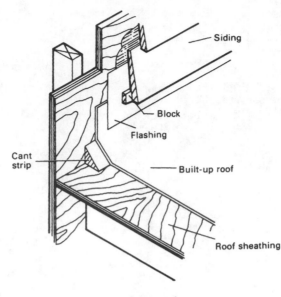

Siding

Block

Flashing

Cant strip

Built-up roof

Roof sheathing

Cant strip

cap The upper member of a COLUMN, PILASTER, DOOR CORNICE, or MOLDING.

capillarity The drawing of a liquid into a porous material by capillary action. For example: masonry in contact with wet ground can draw water into itself. The same is true of wood, which is why it is poor building practice to have any framing members in direct contact with the ground.

cap course Top COURSE of roofing; covers the RIDGE.

cap stone Stone used for the top of a structure.

carpeting Soft synthetic or natural fibrous flooring material.

Carpeting is available in one natural material: wool; and four synthetics: nylon, polyester, acrylic, and polypropylene. Wool is considered the best material. It is soft, luxurious and can keep its look and feel for up to twenty-five years. Of the synthetics, nylon is the best (some 75 percent of all carpet is made with it). It is more stain-resistant than wool (though any stain can be taken off wool if cleaned immediately). Acrylic and polyester are down in the line of quality. They are not as resilient as wool or nylon and are more susceptible to staining. Polypropylene is mainly used for indoor-outdoor carpet.

The pile or thickness of a carpet is a major quality factor. To check this, carpet is made to "grin"—the fiber backing is bent in half to see how dense the fiber is. The less backing that shows, the denser the pile, and the higher the quality.

Manufacturers, of course, work at making carpet seem denser than it really is. The key is the carpet weight per square yard, which will either be on the label or can be asked of the manufacturer.

Padding is also important—it can make a carpet last up to 50 percent longer. But padding must be neither too thick nor too thin, which can have an adverse effect in terms of wearability.

carriage *See* STRINGER.

carriage bolt Bolt that looks like a wood screw and is good for wood-to-wood connections. The carriage bolt differs from the machine screw in that its end is a squarish section under a rounded head with no slot; directly beneath the head is a square shoulder.

Carriage bolts come with either "rolled" or "cut" thread. The cut-thread type has the threads cut right into the bolt; the rolled kind has the threads pressed into the bolt shaft. As a result, the threaded section is a little thicker than the bar shank.

The cut-thread kind is generally preferred. In the smaller bolt sizes, rolled thread works all right; but in larger sizes, the shank or smooth part of the bolt may be loose enough in a large hole large to pass the threaded end. The cut thread will be uniform.

For use, carriage bolts are pushed into a hole the size of the shank, and driven in the rest of the way with a hammer, the squarish portion firmly locking it in place.

Carriage bolt

casement window WINDOW that has a vertical SASH that is opened and closed with a crank handle. Casement windows are available in a variety of stock sizes and in wood, aluminum, vinyl-clad wood, and even pure vinyl. *See* WINDOW.

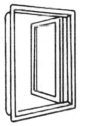

Casement window

casing Rough door framing members. Casing is also used to describe the ARCHITRAVE or molding inside of which a door is hung, but technically it makes up only the inner boards.

cast-in-place CONCRETE that is poured in place as opposed to PRECAST material, which is formed and hardened off-site.

cast iron Material used in the making of WASTE pipes and plumbing FIXTURES.

Cast iron is, in the word of one plumber, the "Cadillac of plumbing." It is extremely strong and rigid and can take tremendous abuse. Moreover, cast iron is a poor conductor of heat so tub water will stay warmer longer in this material than in others. Its only drawback is that it is relatively brittle, but this is not normally important in everyday use.

Making cast iron

Cast iron is made by melting iron at 2700 degrees, then pouring it into a sand mold. When dry it is removed and dry blasted to remove sand particles, sanded, and then etched. It is then coated with porcelain enamel—a combination of clay, feldspars, quartz, and silica stones called "Fritz." (It is applied in powder form and then heated at 1250 degrees, which melts the porcelain into a glass-like coating.

Breaking up cast iron

When cast iron pipes or fixtures have to be removed, they are broken up rather than taken out whole. A cast-iron tub commonly weighs about 500 pounds. It is much easier to don glasses and gloves and rap the material with a small sledgehammer. It breaks up easily into manageable pieces that can then be hauled away.

cat faces Small depressions in the finish coat of PLASTER. They occur because of depressions in the BROWN COAT of plaster, and are corrected when plaster is FLOATED.

cat short 2 × 4s installed between wall STUDS. Cat shorts are used to stiffen and strengthen house WALLS. They are normally installed in a line about halfway up STUD height; if the studs are extra high, there may be more than one line of cat shorts. They also serve as nailing surfaces.

cathedral ceiling A high, vaulted ceiling similar to a church ceiling.

caulk Soft materials used to seal the "seams" of a house. Doors, windows, chimneys, and the like are essentially openings cut in walls and roofing material; they create seams that must be made weathertight. Caulk does this. It is also used inside the house to seal the gaps between tub and wall, sinks and walls, and the like. The essential difference between caulks used inside and outside the house is MILDEW resistance. The former must be mildew resistant, while the latter needn't be.

There is a variety of types to use in building.

- **Aerosol.** This comes in a pressurized can with a nozzle applicator. When dispensed, aerosol caulk looks like white foam, but when it dries, it is hard and waterproof. Aerosol caulk is good for filling openings of various sizes, such as where SIDING meets the FOUNDATION, or where pipes protrude from the masonry.
- **Butyl rubber.** Butyl is an excellent caulk for use with masonry where there is a lot of movement, such as along the joint between the house wall and cement walls or a patio slab. Butyl rubber caulk is difficult to work with. Once a BEAD is laid on, it cannot be tooled without sticking to the applicator.
- **Oil-base.** This is a smooth, dense material available in cartridges and gallon containers of gray, black, and white. It may be used for standard house sealing, but it is particularly favored for "backsealing"—applying a BEAD of caulking to the edges of cedar shingles that abut window frames. The caulk keeps water and drafts from getting under the shingles. It is cleaned up with mineral spirits.
- **Silicone.** This is the best caulk available. Silicone will stick to almost anything, and is relatively easy to work with. There is a type of silicone that will accept paint, but it is quite as good as the standard kind, which does not. It comes in a variety of colors, including clear. *See also* CAULKING CORD.

caulking cord Flat, gray, soft beltlike band segmented into six beads of caulk of various widths. This material is used for setting sink drains (it is less messy than plumber's compound) and for permanent and temporary caulking. It is also used as a base for other caulks where a particularly large and deep gap has to be filled. *See also* CAULK.

cavity wall Two walls with a space between them. Such construction is common in masonry, where the space provides a handy place for any moisture that gets through the outer wall to run out of WEEP HOLES in it, rather than getting through to interior materials and causing damage.

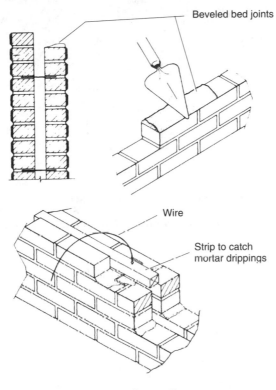

Cavity wall

CCA *See* PRESSURE-TREATED LUMBER.

cedar shakes (roof) Cedar shakes are thick and rough, and are mostly cut by hand. They come in various lengths and thicknesses, depending on grain and size. Long, thin shakes are less likely to warp than wide ones. Shakes cut from HEARTWOOD will stand up better than others.

cedar shakes (siding) Also referred to as "handsplits," these are cut by hand, like CEDAR SHAKES for roofs. The butts range in thicknesses and from 3/8" to 1 3/4" inches. They are available in 18" and 24" lengths.

cedar shingles (roof) Cedar shingles are flat, machine-cut units that range in length from 16 to 24 inches. They are a lovely tan color, highly resistant to decay, and will weather to a gray color.

Shingles are wood, and as such can catch fire. If treated with a fire retardant, they can get a Class C rating, but using them may increase insurance rates. *See illustration on following page.*

cedar shingles (siding) 1. These range in quality. Top is No. 1 Blue Label, which are 100 percent HEARTWOOD and all clear—no blemishes. Just below the blue label is the No. 2 Red Label, which is mostly clear and also looks good. Like other siding materials, cedar shingles go on in COURSES, secured with hot-dipped galvanized nails.

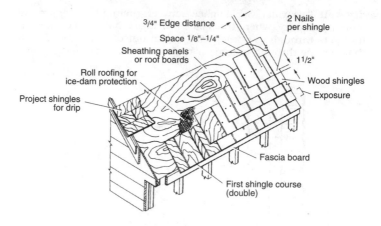

3/4" Edge distance
Space 1/8"–1/4"
Sheathing panels or roof boards
Roll roofing for ice-dam protection
Project shingles for drip

2 Nails per shingle
1 1/2"
Wood shingles
Exposure
Fascia board
First shingle course (double)

Cedar shingles (roof)

2. Flat, machine-cut units with a striated face. Cedar shingles are 18′ long and are ⅜″ or ½″ at the butt. They come in three common grades: perfections nos. 1 and 2, and "undercourse." Perfection no. 1 is the best.

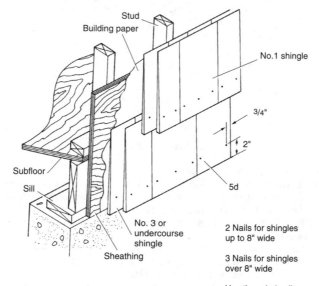

Stud
Building paper
No.1 shingle
3/4"
2"
Subfloor
Sill
5d
No. 3 or undercourse shingle
Sheathing

2 Nails for shingles up to 8" wide

3 Nails for shingles over 8" wide

Use threaded nails for plywood sheathing

Cedar shingles (siding)

cell *See* CORE.

cement 1. In general, glue used to adhere things together. **2.** In masonry, PORTLAND CEMENT, which is used to bind aggregates and sand together to form concrete. *See also* ADHESIVE.

cement board Sheets of hard material made with cement and fiberglass and other waterproof or water-resistant materials.

Cement board, or cement backer board, is used as a backing for ceramic tiles to provide a truly waterproof material as a base. Cement board is waterproof and easy to work with, being installed like drywall. The two most common brand names are Wonder Board and Durock. Wonder Board is almost used generically.

See also BLUEBOARD *and* GREENBOARD.

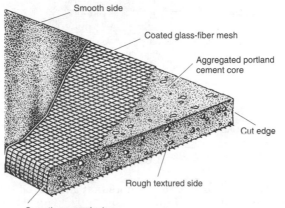

Cement board

centering Temporary formwork used to support MASONRY ARCHES or LINTELS during construction.

centerpiece An ornament in the middle of a ceiling.

centers 1. The distance from the center of one faucet to the center of another. Faucets may be four, six, or eight inches apart. **2.** The distance between STUDS, usually 16″ but sometimes 24″. *See also* O.C. *See illustration on following page.*

> ### Centers didn't always matter
> A couple hundred years ago, carpenters did not worry much about getting studs or other framing members exact distances apart. Modern-day carpenters who work on old homes either know this already or soon will. Old-time carpenters approximated distances, which can be problematic for modern carpenters who work with framing members of inconsistent centers.

central heating system One type of heating system for a building. Central heating systems consists of four different elements: heat producer, exchanger, distributor, and controls.

The heat producer may be an oil burner, gas burner, or, fairly infrequently, an electric heater or coal grate.

The exchanger is what the producer heats: if air is heated, the exchanger is a furnace; if water or steam, a boiler. The exchanger is kept separate from the heat producer as the air, water, or steam is heated.

The heat distributor is the system that, as the name suggests, distributes the heat. There are three different kinds, one for each type of exchanger. In a hot water system, the heated water is driven through pipes by

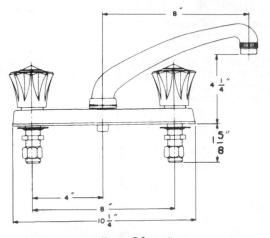

Sinks are usually on 8″ centers.

Centers

a circulator to convertors, or to radiators that give off the heat. In a forced warm air system, the air is drawn into ducts by a large fan and forced out through warm air registers. In a steam heating system, steam rises on its own to heat radiators, which give off the heat.

The other element is the control. Normally this is a thermostat at one convenient location, but it may be several thermostats at various locations for zone control.

ceramic granules The coating on asphalt shingles designed to increase the fire resistance of the shingles and provide good looks. The granules are produced by coloring finely crushed rock under high heat.

ceramic tile Material made of hardened clay and used as flooring and finish wall material.

Where one uses ceramic tile depends on how the tile was made and finished. There is a wide variety of ceramic tiles, but they can be divided into two general kinds: floor and wall tiles.

Most of the tiles covering walls are called FIELD TILES. These are whole or partial tiles that cover a wall completely.

To complete the job, one requires trim pieces. Much like wood trim, these trim pieces are specially formed sections that cover inside and outside corners. Wall tile is available in two basic sizes: 4¼″ × 4¼″ and 6″ × 6″ squares.

Trim is an important consideration in buying tile. Both American- and European-made tile are available, but it is often difficult to get the correct trim pieces for European tile. Without the proper trim, the job cannot be done properly.

Some manufacturers make wall tiles on 12″ × 12″ sheets simulating 4¼″ tiles with grout lines. These sheets do a good simulation job and are easier to install than the smaller units. Wall tile is available in matte and gloss finishes.

Floor tile is made—hardened, chiefly—to withstand foot traffic without scratching or breaking. Two finishes are available: "crystalline" and "clear," which is slightly rippled or a clear finish. To determine how good the tile is, one must learn its ANSI rating. The ratings go from 1 to 5; the higher the number, the better the tile.

Floor tiles come in a number of different sizes: 4″ × 4″ and 6′ × 6′ squares and a 4″ octagon with a "colored dot." The four-inch hexagons are interspersed with 1-inch square, or larger tiles of a contrasting color. Floor tile also comes in mosaic sheets with 1″ × 1″ or 2″ × 2″ squares fastened to a mesh backing. These tiles have an unglazed, flat finish that goes through the entire tile. One- and two-inch hexagons and smaller sizes are also available. The most popular location for the mosaic sheets is the shower, where the surface is valued for its non-skid properties.

Tile is sold according to color lots, which can vary slightly. Experienced contractors always advise that there be sufficient tile from the same lot to complete the job before it is begun.

Wall tile is installed with THIN SET adhesive. It can be installed on ordinary DRYWALL, but there is more protection against water if it is installed on BLUEBOARD, GREENBOARD, water-resistant drywall, or CEMENT BOARD, which is waterproof.

Floor tile may be installed with thin set or with a thick layer of cement, a so-called mud job. In older houses with uneven floors, a mud job often works because it allows the floor to be SCREEDED level, a difficult job if one uses an UNDERLAYMENT like PLYWOOD. Plywood should be used wherever the subflooring is flat so the plywood can be laid flat. If plywood is used, it is important to glue and screw it in place so it cannot move. If it can move, the tile above it can crack.

The final step in installing ceramic tile is applying GROUT.

> ### Most popular—for bath floors
> Nothing is more popular for bathroom floors than ceramic tile. The National Kitchen and Bath Association did a survey that showed that more than half of all baths had ceramic tile for flooring. Vinyl SHEET FLOORING was next, with a little over 42 percent.
>
> Walls were an entirely different story. Tile is used only eight percent of the time with other materials predominating.

cesspool Large BLOCK- or CONCRETE-lined hole in the earth designed to receive and process waste matter from the house. *See also* SEPTIC TANK *and* ABSORPTION FIELD.

CFM Cubic feet per minute. *See also* EXHAUST FAN.

chain bond In MASONRY, a wall held together in part by an iron bar or chain.

chair rail A line of MOLDING that runs around the room at the height of the back of a chair to protect the wall from damage by chair backs. Walls of older homes were frequently panelled, but when decorative plaster came into vogue, it became standard practice to install panelling only about halfway up the wall. (*See* WAINSCOTING.) Eventually, this too disappeared, leaving the chair rail as a remnant of what once was.

chalking Describes PAINT the pigment of which is gradually leaching out. Chalking occurs in many paints as they weather, releasing individual pigment particles. These particles form a fine powder on the paint surface, and are then washed away.

Most paints chalk to some extent. The phenomenon is desirable because it allows the paint surface to be self-cleaning. Chalking is objectionable when it washes over a differently colored surface, or when it causes premature disappearance of the paint film.

Discoloration problems from chalking can be reduced by selecting a paint with slow chalking tendencies. This is usually related to the manner in which the paint is formulated. Therefore, if chalking would be a problem, select a paint that the manufacturer has said is slow-chalking.

Dealing with a chalked surface

Before an excessively chalked surface is repainted, the surface has to be prepared properly to avoid the possibility of peeling. The surface should be scrubbed thoroughly with a detergent solution to remove chalk and dirt, then rinsed thoroughly with clean water and left to dry before repainting.

Chalk that has run down and discolored the surface below may be cleaned by vigorously scrubbing with the detergent solution. If the above problem has been solved, any discoloration will also weather away without scrubbing.

To be on the safe side when repainting, prime the surface with an oil-based PRIMER or a stain-blocking acrylic latex primer. Then apply topcoats as normal.

chamfer Beveled edge of a board. *See also* CANT.

channel 1. Concave groove in a surface for decoration **2.** Decorative MOLDING with a concave groove.

chase 1. Groove or opening in masonry wall for running pipes in. A plumber is said to "chase the pipes through the wall." **2.** Trench for a drainpipe.

check 1. A split in wood. When checks appear in wood, cracks are not far behind. Checking is usually caused by alternating dry and wet conditions. **2.** Small cracks in paint. **3.** Ornamental design composed of inlaid squares.

check rail The upper rail of the bottom sash and the lower rail of the top in a DOUBLE-HUNG WINDOW. Also called meeting rails, check rails are thicker than the rest of the window. The point where they meet is a tight fit for better weather protection. The backs of check rails are usually beveled to ensure this.

chevron 1. Spot where RAFTERS meet at the RIDGE of a GABLE ROOF. **2.** Zigzag pattern used in ornamentation in Romanesque architecture.

chimney General term that describes the structure that contains the components for routing gases and smoke from a fireplace. Strictly speaking, the chimney is the masonry structure that surrounds the other components, but it is commonly used to describe everything. Below is a look at these components and how they work.

- **Foundation.** MASONRY foundation that the chimney rests on. This is also known as a FOOTER.
- **Hearth.** The flat surface directly under the fire. This is the inner hearth. The outer hearth catches sparks that could fly out.
- **Breastwork.** Also known as butt walls, these are the walls that extend from the foundation to the bottom of the flue.
- **Combustion chamber.** This is where the fire occurs. This chamber must be constructed so that the smoke and sparks are directed upwards, towards the flue, and the warmth directed directly into the room. This firebox must be built of fireproof masonry or metal.
- **Damper.** A metal section that can be manipulated, opened, or closed to varying degrees to keep cold air out of the room while providing enough oxygen for combustion.
- **Smoke shelf.** This shelf projects from the breastwork and has a curved surface designed to catch cold air that comes down the chimney and direct it back up and out. The smoke shelf keeps smoke from infiltrating the room, keeping it rising up and out the chimney.
- **Lintel.** Placed across the front of the fireplace, facing the room to create a fireplace opening. It is normally an L-shaped iron section.

- **Smoke chamber.** Wide interior mouth of chimney where gases and smoke collect and travel upward. The smoke chamber works in harmony with the smoke shelf to direct downdrafts of air out of the room.
- **Flue lining.** The flue lining runs from the smoke chamber to a few inches above the top of the chimney. The size of the flue opening is important to allow smoke and gases to escape and the fire to burn properly. The flue lining cannot have any cracks or openings in it, or the fire can escape and ignite any combustible materials near the cracks or openings. For this same reason there must be a clearance about two inches between the chimney and any combustibles.
- **Chimney.** Strictly speaking, this is the masonry walls built around the flue.
- **Chimney top.** The chimney must project a few inches above the flue. Atop the chimney will be a cap to keep rain and snow from collecting there; perhaps a spark arrester, a fine metal mesh to keep sparks from floating up and out onto adjacent roofs; and a hood so rain does not fall directly into the flue. In Europe, large earthenware pots are often used to cap the chimney, but in America chimneys are usually either flush or capped with a stone, although pots can also be used.

Masonry is one option for constructing chimneys. They can also be made of prefabricated metal parts. *See illustration on following page.*

The quest for no fire

In essence, a fireplace allows a fire inside a dwelling to burn safely and without asphyxiating the occupants. Getting to that point has taken awhile.

The first attempt was simply to have an opening in the roof to let out the smoke from a a fire in the center of the room. This did not always work. The smoke was hard to control, and sometimes sparks would set the roof on fire and burn the house down.

One improvement was to build houses from stone and heavy timbers that would not readily burn down. While this cut down on fires, there was still no way to control the smoke, which still had to exit the building via a hole in the roof. The hole let the smoke out, but it allowed rain, drafts, dirt, and other things in.

Another advancement was a cover on the roof that let the smoke escape but also kept out rain. The problem was still the smoke: it remained uncontrollable.

The breakthrough came in the fourteenth or fifteen century, when someone designed a tall pipe to route the smoke up and out. The first chimneys were not, by any means, perfect. Smoke was still something of a problem, and houses would occasionally still burn down.

Over the years, chimney design has become better and better. Today, one can sit in front of a crackling fire and not have to worry about smoke or a fire.

chimney blocks Vertically stacked CONCRETE BLOCKS that create a FLUE.

china *See* VITREOUS CHINA.

chord Any of the outside members of a TRUSS connected and braced by web members.

chrome-plated Bath items—plastic as well as steel—commonly come chrome-plated. Chrome resists corrosion and water very well.

cinder block Building block whose AGGREGATE is cinders. Years ago, lightweight blocks were made with cinder aggregates, but this is hardly the case anymore. Nevertheless, the name is commonly used to describe all kinds of block, including CONCRETE BLOCK.

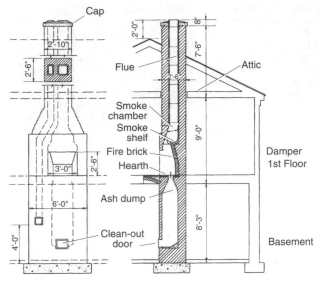

Chimney

circuit The series of wires that electricity travels on. There is a variety of circuits in the average home: general purpose, small-appliance, and individual appliance.

General purpose circuits, also known as *lighting circuits,* power items that do not draw a great amount of power, such as lights, TV sets, small fans, vacuums, hi-fi equipment, and the like.

A house designer will make every effort to include enough circuits not only for an occupant's present needs, but also for possible future expansion. There are various ways to figure what circuits to include. One way is by square footage. It is generally accepted that there should be one branch circuit for every 500 square feet of area in a house. Wattage is another way. It is accepted that circuits should provide three watts of power for every square foot of living area. To determine the total number of circuits, the total wattage is divided by 1500, the wattage that a standard circuit should be able to carry.

The number of OUTLETS in a house is also important to calculate. A minimum is usually figured at one "duplex" RECEPTACLE for every twelve feet of linear footage in a house, or a portion of this. For example, three outlets would be required for 29 feet: two for the first 24 feet, and one for the five additional feet.

Small appliance circuits, as the name suggests, provide power for small appliances. They should be able to carry up to 20 AMPS and Number 12 wire should be used. Deep fryers, blenders, toasters, and the like will all run off small appliance circuits. The circuits should be located wherever the power will be consumed, such as the kitchen or den.

The NATIONAL ELECTRIC CODE suggests that a house have two such circuits, each with a capacity of 2400 VA. Some small appliances have very high wattage demands. A roaster, for example, may use 1500 WATTS.

Individual appliance circuits are for heavy power users such as clothes dryers, water heaters, garbage disposer, and dishwashers. It is customary to provide individual circuits for these things. Circuits may range from 20 to 50 amps (for an electric heating system). Although they don't have large power needs, furnaces and boilers are also earmarked for individual circuits to isolate them electrically so they will be unaffected if other circuits go down. *See also* CIRCUIT BREAKER *and* ELECTRIC SERVICE.

circuit breaker A safety device that turns off if an electrical circuit malfunctions.

Circuit breakers are the modern equivalent of the FUSE. Indeed, they are much more popular, though many electricians feel fuses are safer.

Circuit breakers are popular because of their convenience. A flip or push of the switch gets a circuit going again, without having to unscrew a fuse and screw in a new one.

A circuit breaker senses overloads and trips out in order to avoid overheating a wire and possibly causing a fire. At the heart of its construction is a bimetallic strip, two springs, and contact points. When a circuit overloads or a SHORT CIRCUIT occurs, this strip bends and pulls away from the contact points, breaking the circuit and shutting the electrical flow down. When the breaker lever is reset—flipped—the strip is pushed against the contact points and the electrical flow starts again.

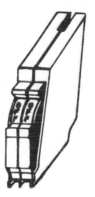

Circuit breaker

cistern 1. Receptacle for water storage in older houses. **2.** Another name for the WATER CLOSET in old-fashioned toilets.

clapboard A type of wood SIDING that consists of long beveled boards.

Clapboard is today a general term used to describe a variety of individually styled siding that share only the essential characteristic of being long and beveled in cross section.

When installing clapboard, or any bevel siding the exposed "to the weather" portion of the boards should not be less than one inch. The average exposure is usually determined by the distance from the underside of the window SILL to the top of the DRIP cap. For weather resistance and appearance, the BUTT edge of the COURSE is installed so that it sits on top of the drip cap. In many one-story houses with an overhang, this course of siding is replaced by a FRIEZE board. It is good for the bottom of the course to be FLUSH with the underside of the window sill, but this is not always possible, because of varying window heights and types.

The installation of clapboard begins at the bottom of the house, like other types. The first course is normally blocked out with a "starting board" of the same thickness as the top of the board. Each succeeding course should overlap the upper edge of the lower course, and should be nailed to STUDS with 1½" minimum penetration into the stud. If backer board or the like is used, nail length should be adjusted to maintain the right penetration.

Nails go through the bottoms of boards and are placed so that they are about ⅛″ above the board below—they miss it. This allows the boards to expand and contract from weather changes without splitting.

Installers try to avoid BUTT JOINTS wherever possible. Longer sections of siding are used under windows and long stretches. Shorter lengths of siding are used between doors and windows. Where butt joints are not avoidable they would ideally be made over studs and staggered between courses as much as possible.

There are also a number of vinyl sidings (*see* SIDING) that imitate clapboard.

The first clapboard

Clapboards are an American creation, having their origin in Colonial days, but where the term comes from is debatable. There are various explanations, including one that the word derives from cloveboards, which referred to the way the first clapboard were made. Oak logs were set on end and then split with a tool called a froe, or riven into the boards. Later, saws were used, allowing longer and smoother board to be made.

Originally, there was no SHEATHING on early buildings. Overlapping clapboards were simply nailed to studs, making for a drafty, cold interior. Board sheathing was introduced by the 1800s.

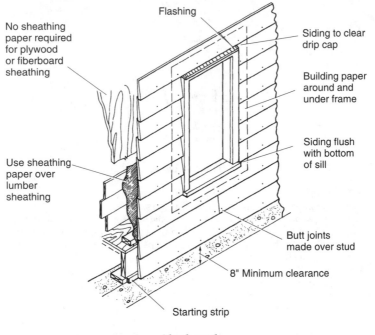

Clapboard

class A A fire-protection rating for wood shingles and roofing.

cleanout 1. In plumbing, the main TRAP opening at the bottom of the DWV system. **2.** In MASONRY, an opening in the bottom of the FORM for the removal of waste material. **3.** In masonry, an opening in the first course of a reinforced concrete wall for removal of mortar protrusions and drippings. *See illustration on following page.*

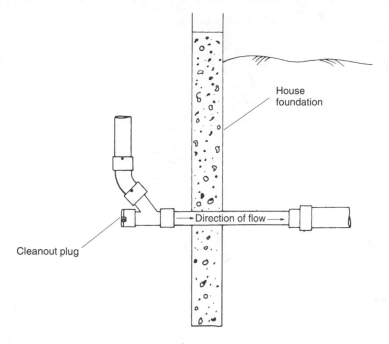

House
foundation

←Direction of flow→

Cleanout plug

Cleanout

clear A grade in lumber (BOARDS) that describes the material as being clear of knots.

cleat A piece of wood that acts as a support. Cleats can be used to help support cabinets, shelves, and framing members, to name a few.

clip Cut-to-length BRICK section.

closed cornice *See* CORNICE.

closer **1.** Space to be filled by a short length of BRICK. In any given brick construction, spaces are created that require bricks shorter than full length. Such gap-fillers are called closers, or closures. **2.** Final brick required to complete building a brick wall.

closet Storage room. *Closet* comes from the French word for "little room," and it is actually that.

closet bend Plumbing FITTING that connects the WATER CLOSET to the soil STACK. The closet bend is a straight piece of pipe with one end bent. The straight portion connects with the soil stack while the bent end connects with the W.C. *See illustration on following page.*

coal tar pitch Brown or black semi-cold hydrocarbon formed as a residue from the partial evaporation or distillation of coal tar. Coal tar pitch is used as waterproofer in dead-level or slightly sloped built-up roofs.

code *See* BUILDING CODES.

cold process roofing Semi-flexible asphalt roof covering applied with roof cement.

cold weather concreting Pouring concrete when the temperature is below freezing. The danger is always that the water used in the mix will freeze, damaging the material. Special ADMIXTURES are added to make this less likely. *See* AIR ENTRAINED CONCRETE.

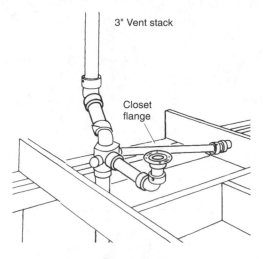

3" Vent stack

Closet
flange

Closet bend

collar beam A reinforcing wood member fastened between rafters on opposites sides of a roof.

Collar beams are normally fastened a foot above wall PLATES. If the beams are connected at the lower ends of rafters, they are known as tie beams. Collar beams probably get their name from their location, roughly where one would expect the collar to be if a house were a person.

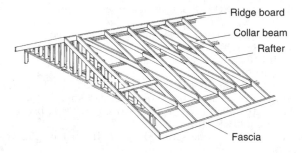

Ridge board

Collar beam

Rafter

Fascia

Collar beam

collar joint The long vertical MORTAR joint between WYTHES of masonry.

collar ties *See* COLLAR BEAM.

column A pillar, usually round, which acts as a support for a structure above. Columns consist of a base on which the pillar rests, the pillar or shaft, and the top or head of the column, known as the capital.

common lumber BOARDS with a number of imperfections and knots; commonly used by builders for construction in areas where good looks is not the major factor.

compaction Process of PUDDLING or otherwise vibrating fresh concrete so that all air pockets are eliminated and it hardens without flaws.

compass brick Curved brick used in constructing an ARCH.

composite panel Brand name for a veneer-faced PLYWOOD panel with a reconstituted wood CORE.

composite wall Collection of MASONRY walls where the masonry materials used differ. *See also* WALL.

compressive strength The ability of a material to withstand downward force or weight.

concave A line that follows the interior of a sphere. People often confuse this with convex, and vice versa.

concave joint Joint made in water-resistant MORTAR.

concrete A mixture of cement, sand, aggregate, and water that produces a hard, durable material for all kinds of building.

Concrete cures by hydration, chemical- (as opposed to air-) drying. While concrete is drying, it must be kept damp for a week or so to permit it to dry gradually. Concrete will not be at full strength for a month or so.

While a wonderful building material, moist concrete will eventually develop some cracks, so it is important to take waterproofing steps. Before finishing concrete walls, for example, it is important they be waterproofed. Although this can can be done with tar, there are a number of other products that do equally well at waterproofing. One essentially consists of cement and rubber. Applied to raw masonry, it penetrates the pores with a grip so tenacious that even substantial water pressure will not drive it off.

Similarly, DRAIN TILES should be used around basement walls.

Concrete SLABS can crack as well. A good precautionary measure to prevent water intrusion is to apply six mil polyethylene to the ground before pouring the slab.

concrete blocks See BLOCK.

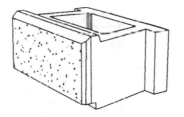

Concrete block

concrete block segmental retaining wall system Mortarless concrete BLOCK used in making RETAINING WALLS.

These blocks are made in such a way that they interlock, and are held together with fiberglass pins or the like. Their weight, plus the "batter" or setback or incline, can keep a wall that is up to 5½ feet from being toppled. Over that height, engineers say that the blocks can be used, but they require additional support, grout, or special HARDWARE to keep them erect. Both straight and curved walls may be built.

These block, which come in a few different colors, handle like regular block, but they are heavy, weighing about 70 pounds each for professional use. There are units for do-it-yourselfers that weigh 20 to 25 pounds. The block can be cut with diamond tipped saws. *See illustration on following page.*

concrete brick solid Block made of PORTLAND CEMENT, water, and aggregate. Such blocks are often referred to by the type of aggregate used: concrete, cinder, lightweight, etc.

concrete masonry unit MASONRY unit usually not larger than 4″ × 4″ × 12″.

condensation Moisture created when warm, moisture-laden air contacts a cold surface. *See also* DEWPOINT.

Condensation can create a host of problems in just about every area of the home. It can rot timbers, peel paint and overall, make a domicile cold and clammy. To understand it is to be able to deal with it effectively.

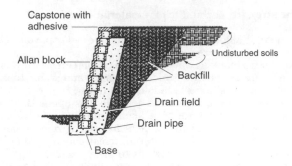

Capstone with
adhesive

Allan block

Undisturbed soils

Backfill

Drain field

Drain pipe

Base

Concrete block segmental retaining wall system

Water vapor is a gas with molecules so tiny that they can pass through wood, masonry, and anything else that has some permeability to it. Inside the house, water vapor is constantly being generated. This vapor will seek to pass to an area where the air contains less moisture. Warm air can hold more water vapor than cold air, so when the house is warm and the outside air is cold, the moisture-laden air transfers, assuming nothing is blocking it, to the drier air.

The problem is that this warm air can contact a cold surface, resulting in condensation.

> **Condensation, byproduct of modern times**
>
> Condensation has only become a pronounced problem in the USA since the end of World War II. Until that time, most homes were uninsulated, so moisture-laden air could freely pass through walls out of the building without causing problems. When insulation came along, this moisture vapor was blocked to some degree, creating condensation problems.
>
> Adding to the amount of moisture in the air was the emergence of devices—washer, dryer, and other appliances that generate it.

conduction A "mechanism" of heat flow. The mechanism can be seen in the way heat is transferred in the metal handle of a frying pan. As the pan is heated, the handle becomes hot, exclusively the process of conduction.

With most materials, the denser the material, the higher its rate of heat flow by conduction. Metals transfer a great deal of heat this way, and so they are commonly used in electric transmission systems. Because they are excellent conductors, energy loss is at a minimum. Conversely, metal elements that extend through building sections are undesirable because of their high level of conduction. *See also* CONVECTION.

conductor Electric CABLE. *See also* WIRING.

conduit In general, a conduit is a hollow tube or pipe that comes in various forms and materials and through which electrical wire is run.

A Greenfield looks just like BX, except there are no wires in the conduit. They are pulled through after the Greenfield is in place. The technical name for this is "flexible metal conduit." Greenfields are designed for indoor use, but they can be used out-of-doors with wiring that has a W in its name, such as TW.

Greenfields are normally not used for entire wiring systems, with a few exceptions. They are normally used inside, in a permanently dry location and where some movement is required of the conduit, such as a motor. In certain situations Greenfield can serve as its own GROUNDING wire, and in others it needs a separate wire.

A heavy wall conduit is designed to be used wherever there is danger of physical abuse, such as if buried in a lawn. Another common use is carrying wires from a house to a garage. Heavy wall is used in new construction.

The pipe can be larger than necessary, and then new wires pulled in as needed later. It is secured to electrical BOXES with nuts.

Rigid conduit is cut with pipe cutter and bent with an electrical bender. It is too laborious a job with manual tools. A variety of fittings are available to allow it to make turns. Rigid conduit is difficult to work with because it must be cut to fit precisely between boxes—just long enough to screw into them.

Thin-walled pipe (1/16" thick) is designed to carry electrical wires, usually inside the house. It may also be used outside. This conduit, also called EMT (for electric metallic tubing) comes in internal diameters of 1/2" to 4" and even larger and in 10 foot lengths. The 1/2" diameter is most commonly used in homes. EMT (up to 2" in diameter) is too thin to thread, so a variety of pressure fittings are available to make connections.

See also ELECTRIC SERVICE.

Heavy wall conduit

Greenfield

consistency A general term used to describe the workability of a wide variety of building materials such as PLASTER, ADHESIVE, JOINT COMPOUND, and CONCRETE. Proper consistency is key to a successful building job. If plaster is too thin, or concrete too thick, there is no way to a successful job.

construction adhesive General term for ADHESIVES used for various building tasks.

A wide variety of adhesives for building are available. They are available in tubes, in cartridge form for dispensing through a cartridge gun, and two-gallon and larger containers. A number of caulks, chiefly silicone, also serve as excellent adhesives.

Construction adhesives are normally meant to supplement mechanical fasteners such as nails. For example, paneling adhesive is applied in a wiggly bead to the backs of panels, but the panels are also nailed in place. Weatherproof adhesive is used to help secure decking boards, but nails are also used. Adhesives give an extra measure of bonding strength, ensuring that the material will stay in place tightly—movement can lead to installation failure.

construction drawings Describes the various kinds of drawings that the builder must be able to read in order to build. Construction drawings include architectural plans, blueprints, whiteprints, building plans, and working drawings. They all show one thing: how to construct the building. A couple of concepts should be understood to work with construction drawings.

All drawings are to scale. A fraction of an inch on the drawing stands for a foot in the actual construction. On smaller buildings such as homes, the normal ratio is 1/2" to one foot. This facilitates finding the length of any part of the home. Just measure the inches on the drawing, and multiply it by the ratio. For the home mentioned earlier in this paragraph, a three-inch line represents a twelve-foot distance in the completed home.

The thickness and style of lines are used to indicate various aspects of the structure. In general, the thicker the line, the more significant the part of the structure that it represents.

For small buildings, there are generally three types of drawings: site plan, floor plan, and "elevations." Occasionally detail and sectional drawings are also needed.

The site plans show where things are to be, including the exact dimensions of the finished building, and where things are. It looks straight down on the property. On it, one will see locations of incoming water, gas, and electric lines, the boundaries of the building, and potential problems such as trees. The site plan also reflects any walkways or other buildings and the exact position of the foundation. A curved line on the site plan indicates the slope of the land, if any. Like other drawings, the one for the site plan will have to be shown to the BUILDING CODE authority for approval.

The floor plan, also shown from the point of view of looking straight down, shows the rooms and their sizes; where doors are, and which way they swing; where windows are; and the location of such things as fireplaces and the dishwasher, stove, refrigerator, and more.

Elevation drawings show the building from the side—actually all four sides—and indicate the height of things that could not be listed on the other drawings.

It is sometimes necessary to take a closer look at something on a drawing, and this is the purpose of the detail drawings. For example, the architect may have framed out the corners of the building in such a way to resist WIND LOADS. The drawing is enlarged so that one can view details clearly.

A section drawing is a cutaway view of a section of a building. If one wanted to show how door framing was, the architect or builder could cut the area in half and provide a view from the side.

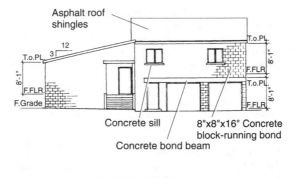

Concrete sill   8"x8"x16" Concrete block-running bond
Concrete bond beam

Front / East Elevation
1/4" = 1'–0"

Construction drawing

contact cement Adhesive that bonds on contact.

Contact cement is commonly used to BOND PLASTIC LAMINATE to kitchen cabinets and COUNTERTOPS. It is applied to both surfaces to be joined and allowed to dry to the touch. The two surfaces are carefully brought together—carefully because they will bind together at the point that they meet with no realignment possible.

Contact cement comes in water-base as well as solvent-based types, the latter in low- as well as high-FLASHPOINT forms. The high-flashpoint type is used where there is danger of explosion (the vapors are flammable). Most installers do not have high regard for water-based contact cement. There is also a nonflammable solvent-based type that cleans up with ammonia and water that is said to be as effective as the solvent-based version.

continuous beam Framing member that rests on more than two supporting members of a structure.

control joint **1.** Joint that penetrates only partially through concrete slab. If pressure, such as FROST HEAVING, is applied to concrete, the control joint is designed to crack, rather than the concrete. **2.** In CONCRETE BLOCK construction, a continuous vertical joint designed to relieve expansion stress. Such a joint can be made with special blocks, or by installing a strip of building paper on one side of the MORTAR JOINT. Such joints are normally put in any of the different spots where concrete-block walls are most likely to be weakest: where the height of the wall changes; next to or above doors and windows and CHASES; at the point where a PILASTER meets with a wall; where two walls meet.

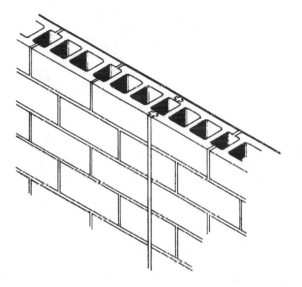

Control joint

convection A mechanism of heat transfer; one way that heat flows.

In general, heat flows from hot to cold areas (*see* HEAT TRANSFER). Convection is one way this is done. Heat flow by convection can be seen in a two-story house where the second floor is warmer than the first. The difference in temperature occurs because heated air is less dense than cooler air. The heated air moves up and across the ceilings of the first-story rooms, and up the stairway to the second-story rooms. At the same time, the denser cooler air settles to the floors of the upper-story rooms, and moves across the floors and down the stairway to the first-story rooms. *See also* CONDUCTION.

coped joint Joint cut with a coping saw and used on baseboards, doors, window stops, and other moldings.

coping **1.** In MASONRY, the top layer of stone or other material designed to protect the masonry from water penetration. Coping is usually made of sloped STONE or other slanted MASONRY product so it readily sheds water. **2.** The process of making a COPED JOINT in MOLDING. *See illustration on following page.*

copper pipe *See* WATER PIPE.

core **1.** Hollow section of BLOCK. **2.** Center section of a DOOR. **3.** Center section of a PLYWOOD panel.

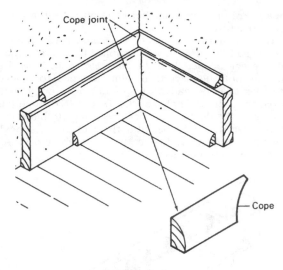

Coping

corner bead Metal molding installed on corners of a DRYWALL job. Corner bead protects the corners against chipping.

corner block *See* BLOCK.

corner board Boards installed along corners of a house to protect corners and ends of siding.

corner bracing Diagonal brace placed at the corner of a frame structure designed to stiffen and strengthen the wall.

cornice Lower section of a roof that forms an overhang. A cornice, also called an EAVE, is collectively composed of a FASCIA, SOFFIT, and soffit MOLDING. In GABLE roofs, the cornice is formed on the long sides of the house. With HIP ROOFS, it is continuous around all sides.

In a narrow-box cornice, the soffit board is narrower than on other types of cornices. The soffit is nailed to the bottom of RAFTERS. The truss roof version has a small horizontal return wedge. Though narrow, there is more than enough space in the soffit boards to install attic ventilators. Because it is narrow, a narrow box cornice does not provide shade, nor allow sun to heat passively in the winter when it is low in the sky. *See illustration on following page.*

A wide-box cornice normally requires another horizontal member, attached to each truss, to which the soffit is nailed. Trusses can be ordered with returns attached (wide-box cornice with returns). When rafters are used, LOOKOUTS are toenailed to the wall and FACENAILED to the end of the rafter overhang. *See illustration on following page.*

A wide-box cornice without returns is sometimes used for houses with wide overhangs. The soffit material is nailed directly to the underside of the rafter extensions. Inlet ventilators are installed in the soffit area.

An open cornice is structurally the same as a box cornice without returns or lookouts, but the soffit is eliminated. Open cornices are often used in POST-AND-BEAM construction with large, widely spaced rafters and with 2 × 4 and 2 × 6 DECKING used at the overhanging section.

The open cornice requires that the blocking be toenailed in place between rafters or trusses to close the space between the top of the wall and the bottom of the roof decking. If trim is desired, blocking is best placed vertically. The trim board must be carefully notched to fit around

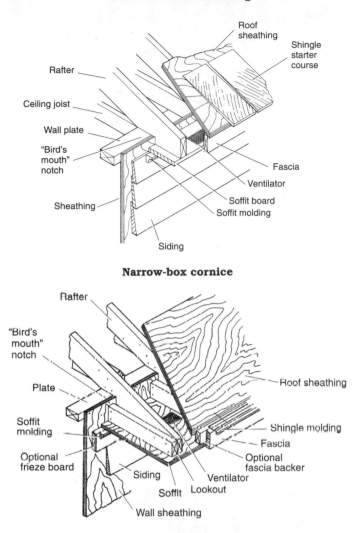

Narrow-box cornice

Wide-box cornice

the rafters. Roofing nails protruding through the exposed decking can be clipped with snips, and a higher grade roof decking can be used around the perimeter of the roof to enhance the appearance of the underside of the overhang.

A closed cornice is, in effect, no cornice at all. There is no soffit board, but the wall SHEATHING or exterior panel siding extends up to the rafters or trusses. The ends of the roof are with or without a fascia board, or a simple molding. Closed cornices are, of course, simple to build. On the other hand, they do not provide any protection on the side of the house, nor do they look good. Appearance can be improved by adding gutter.

A cornice return is a cornice that continues around the entire house. The roof design must allow this, as a hip roof does. *See illustration on following page.*

counterflashing Flashing used on chimneys at the roofline to cover shingle flashing and to prevent water entry. *See illustration on following page.*

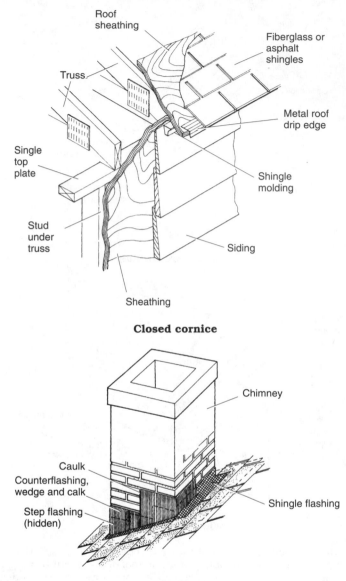

Roof
sheathing

Fiberglass or
asphalt
shingles

Truss

Metal roof
drip edge

Single
top
plate

Shingle
molding

Stud
under
truss

Siding

Sheathing

Closed cornice

Chimney

Caulk

Counterflashing,
wedge and calk

Shingle flashing

Step flashing
(hidden)

Counterflashing

countersink Drive a nail or screw beneath the surface, filling the resultant depression with wood putty or a plug. Of the two methods, the plug will undoubtedly work better because it is difficult to get a patching material stained to match the wood that is there. (If the surface is being painted, this is not a problem.) *See illustration on following page.*

countertop Unit that rides on and is secured to base cabinets in a kitchen. The standard countertop is built to fit a particular space and covered with PLASTIC LAMINATE. Others are post-formed.

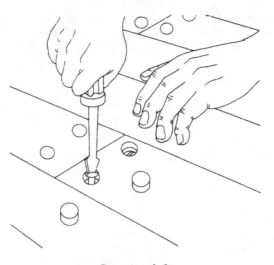

Countersink

The countertop is installed after the BASE CABINETS. After locations for OUTLETS are marked, the countertop is set on top of the cabinets and the outlet marks transferred to the BACKSPLASH. A jigsaw or other power saw cuts out the necessary openings from the back side to reduce the chances of chipping the laminate. The countertop is then set back in place. If it does not fit snugly, the DRYWALL or PLASTER is cut to make sure it does.

Once the countertop fits, it is ready to be fastened to the cabinets. This is done by driving screws up through the base cabinets and into the countertop. As the screws draw tight, the countertop is drawn tightly into position on the cabinets.

The post-formed countertop is made in the standard way, but it is covered with laminate heated to follow the contours of the material, creating a piece that is seamless, except at the sides. Post-formed countertops come in a variety of lengths, colors, and designs. Installing one is much like installing a standard countertop. A sink, cooktop, or the like is used as a template, and the openings in the top are made from the back side of the top to avoid chipping the laminate. The ends are handled differently: BATTENS are nailed and glued to the underside of the countertop near the edge to provide bulk for edge strips of laminate. These strips have adhesive that is activated by an iron. In other words, the laminate is laid in position on the edge, and then an iron, set to medium heat, is run along it to activate the adhesive.

The countertop is then laid in place, checked for level, SHIMMED as needed, and screwed down tight from the underside as with a standard countertop. If the countertop consists of more than one section, the sections are joined with siliconized latex caulk and clamped together until the caulk sets up.

Fitting the counterop to the wall is essentially the same as for any countertop. The countertop is pressed against the wall and the wall material removed to accommodate the top. A compass is used to trace the wall outline on the backsplash, which is trimmed with a saw.

coupling In plumbing, a type of FITTING that is used to join pipe sections.

course A continuous horizontal line of installed material. *See* BRICK, BLOCK, DAMP COURSE, *and* SIDING.

cove Molding with a concave face used to finish interior corners. *See* MOLDING.

coverage In general, the area a particular material can cover. One might speak of paint covering 400 square feet per gallon, roofing covering 100 square feet per SQUARE, or a bundle of hardwood flooring covering square feet.

CPVC Chlorinated Polyvinyl Chloride, a type of plastic pipe used to carry hot water. *See* WATER SUPPLY SYSTEM.

cradling Framework used to support lath and plaster on a vaulted ceiling.

crawl space Open space under a basementless house. The crawl space is normally surrounded by the FOUNDATION wall. Crawl space is so named because it is too low to walk around in; crawling on hands and knees is about all one can do. Its function is to allow inspection of pipes and ducts.

Crawl spaces are built, in part, because in many areas of the country they are more economical to build than other types of foundations. They do not require the extensive EXCAVATION or grading, except for FOOTINGS and walls. In mild climates, the footings are installed only slightly below finish GRADE. The exception would be in cold northern states and Canada, where frost penetrates deeply and the footing can be four feet or more below finished grade. In this case, it would be almost economical to build a full foundation and get a lot more space.

There are, generally, two types of crawl spaces: wood and masonry.

PRESSURE-TREATED LUMBER is used in the same manner as it is when constructing a full FOUNDATION, except that the plywood panels used as SHEATHING only need to be two feet high. The footers can also be wood, and the studs only need to be on 24" centers instead of 16". Excavation need only go as deep as the crawl space floor.

If local frost conditions require greater depth, a trench of appropriate width is dug around the perimeter, allowing the wall to extend down to the required depth. A layer of crushed stone or gravel with a minimum depth of 4 inches is deposited at the bottom of the trench and leveled. Wall panels are installed over footers placed on the gravel and braced in place, plywood joints are caulked, and the wall is covered with 6 mill polyethylene below grade on the outside. If the SPAN warrants it, a wood-frame center bearing wall may be built. This can be built from 2 × 4's spaced 24" O.C.; plywood facing would not be required.

A MASONRY wall for a crawl space is like a foundation wall but with some important variations. No excavation is required within the walls, and masonry piers replace the wood or steel posts used to support the center beam of the basement house. FOOTING size and wall thickness vary with location and soil conditions. A common minimum thickness for walls in a single-story frame house is 8" for hollow concrete BLOCK and 6" for poured concrete. Minimum footing thickness is 6", width is 12" for concrete block, and 10" for poured concrete.

Poured concrete and concrete block PIERS are often used to support floor beams. They should extend at least 12" from groundline. Minimum size for a concrete block pier should be 8" × 16" with a 16" × 24" concrete footing that is 8" thick. A solid block cap is used as a top course. Poured concrete piers would be at least 10" × 10" with a 20" × 20" footing that is 8" thick.

Non-reinforced concrete piers should be no higher than ten times their smallest cross-dimensional dimension. Concrete block piers should be no higher than four times the narrowest cross-sectional dimension. Spacing of piers should not exceed eight feet on centers under exterior wall beams set parallel to floor JOISTS.

Exterior wall piers should not rise above grade more than four times their smallest dimension unless they are supported laterally by masonry or concrete walls. The size of the pier for wall footings should be based on the load and bearing capacity of the soil. *(See illustration on following page.)*

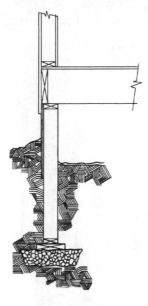

Crawl space

crazing 1. A network of fine cracks in a finished paint surface. This condition can be caused by a variety of things, but it will eventually peel. **2.** Fine line cracks in an asphalt surface.

creep Deformation of a built-up roof system caused by thermal stresses.

cricket A small tent-shaped construction, also called a saddle, installed behind a chimney on a sloping roof. The purpose of the cricket is to route water so it does not collect behind the chimney.

The origin of the term is unclear. One Georgia Pacific architect opines that it might have to do with the sharp angle that a cricket's legs are at—basically a V shape—when it is at rest.

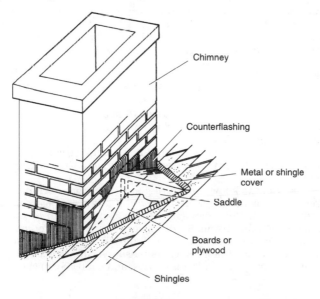

Cricket

crimp A crimp formed in sheet metal for fastening purposes or to make the material less flexible.

cripple 1. Framing member is cut shorter than usual. Cripple STUDS, which are used in framing out windows, commonly utilize cripples. Cripple rafters are also used.

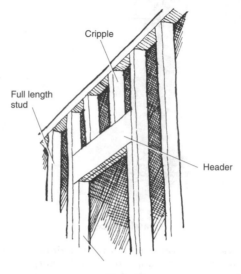

Cripple

crossband In PLYWOOD, veneer layers with GRAIN direction perpendicular to that of the face plies.

cross grain *See* GRAIN.

cross joint Joint between the ends of two MASONRY units.

crowding the line In MASONRY, laying masonry units so they touch the mason's guideline. *See also* MORTAR.

crown molding A MOLDING used on a cornice or wherever an interior angle is to be covered. If a molding has a concave face, it is called a convex molding.

culls BRICK that has been discarded for being below manufacturing standards.

cup Crosswise distortion of a structural wood panel from its flat plane. *See also* BOW.

cupshake *See* WINDSHAKE.

cupola Dome-like structure on the top of a building. A cupola, which comes from the Latin for "little barrel," may be used as a lookout, a light house, or for aesthetics. The word "cup" derives from cupola.

cupping the mortar In BLOCK laying, cutting and rolling the mortar with a trowel off the mortar board.

curing The process by which a material reaches full strength. Many materials, including PLASTER, certain adhesives, paint, and CONCRETE, require curing times.

It is important to know what curing times are. For one thing, curing relates to strength. An item may seem firm, but if it is used before curing and full strength occur, the item may fail. Curing also relates to the acceptability of finishing. One should wait 60 days, for example, before coating concrete. Why? Alkalis in uncured concrete will interfere with adhesion. The same is true of plaster. Some things, such as masonry, can cure for years. The water in the masonry very gradually dissipates, and subtle chemical changes occur.

current Electrical current, or flow. Electrical current is referred to in terms of the amount that flows, or amperage, measured in AMPERES. Devices that use the current use power or wattage, which is measured in WATTS and is determined by multiplying amperage by VOLTAGE. Current is also either AC or DC.

Electrically, it's not the same

Electricity is standardized in the United States, but this can hardly be said of the rest of the world. When electricity was in its infancy, manufacturers designed systems where only their particular products could be used. Many an American traveler has been frustrated when confronted by a receptacle, for example, that his appliance doesn't fit into.

The differences go well beyond this. Voltages throughout the world may range from 110 to 440. France offers 120-volt service, but also 220, 240, 380—and many more. Turkey provides power up to 380 volts, and in Hong Kong it is lower—only 346. The alternating current standard in America is unknown in many parts of the world. While most use alternating current, it is at a frequency of 50 rather than 60 CYCLES.

Using American products can range from inconvenient to disastrous. Some American devices, such as an electric razor, can run on 50-cycle AC; they will just run more slowly. But a clock set on that current will lose time. And if the motor of the particular device is set for 120 volts and gets zapped with an unexpected 220, you'll be looking for a new appliance.

curtain wall 1. In masonry, a variation on a VENEER WALL. Instead of veneer tied to a backup wall, the curtain wall is tied to a framework of metal framing members and encloses the space behind the framing: it runs all the way from the foundation to the roof. Another wall could be added to make a CAVITY WALL, and it could also be built as a COMPOSITE WALL. Curtain walls are tied to the FOUNDATION, to receive its full support.

Since curtain walls are secured to concrete or steel frames, they will expand and contract differently than the frames, so it is important that the TIES used are flexible, and made of GALVANIZED STEEL or other non-corroding material.

2. In a wood-framed house, it performs the same function as a PARTITION WALL.

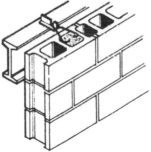

Curtain wall

custom cabinets Kitchen cabinets constructed to fit a particular space. *See also* KITCHEN CABINET.

cutback COLD PROCESS ROOFING adhesive or similar material that has been thinned with solvent.

cycle *See* AC.

dado 1. Woodworking JOINT consisting of a slot and a corresponding section that fits into the slot. 2. The lower half or partial section of a wall. This section is often panelled. The word derives from the Italian word "die," the part of a pedestal between the base and the CORNICE.

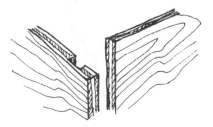

Dado joint

damp course COURSE of MASONRY material on the ground that prevents moisture penetration from the ground due to capillary action. Also known as a "damp check."

dampproofing In MASONRY, prevention of moisture penetration by capillary action.

DC Direct current. Electric current that flows in one direction, rather than alternating. A car battery is typical of direct current. *See also* AC.

deadbolt lock In a deadbolt, when the latch on the lock slides into the "strike" or slot in the DOOR, it cannot be pried; a doorknob or latch has to be turned.

The double-cylinder deadlock opens and closes with a key from the out-side and the inside. A double-cylinder deadlock is an excellent security lock. Even if a door has glass panels, a potential burglar cannot break a pane of glass and then reach in and turn the knob; it can only be released by key. It can also frustrate a burglar who has broken into a house and then encounters a double-cylinder type in the back, which will delay and perhaps compromise his escape. It should be noted that this type of door lock can be dangerous in the event of a fire because it will need to be opened with a key in a smoky, high-stress environment.

Double cylinder deadlock

A mortise is a squarish box that does not have knobs. Installed inside the door and difficult to pick, it is the ultimate kind of lock for security.

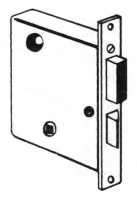

Mortise deadbolt

A night latch is a squarish box with a prominent latch. This is a keyless lock. When the door slams, the night latch locks. It can be opened by turn-ing a knob. It is surface-mounted.

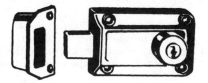

Deadbolt lock: night latch

A rim is a rectangular lock with one rounded end and rings through which the strike bolt slips when closed. It also has a knob inside the lock that lifts the strike in and out of position. A rim lock is similar to a night latch in that it has no key, and it is considered jimmy-proof.

Deadbolt locks are normally for exterior use.

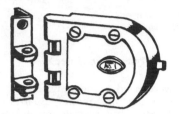

Deadbolt rim lock

dead load The weight, expressed in pounds per square foot, of elements that are part of a structure. Dead loads include all of the weight of all of the framing members, the roofing, drywall, plumbing, heating, and any other "equipment" that could be considered a part of the permanent weight of the house.

Engineers and architects use dead load figures to calculate just how strong the framework must be. *See also* LIVE LOAD *and* WIND LOAD.

deadman timber A large buried timber used as an anchor for a RETAINING WALL.

decay *See* DRY ROT.

deck 1. An elevated wood structure in the back of a home. 2. The wood base for ROOFING, also known as SHEATHING.

PLYWOOD is the most commonly used roofing material. How thick the material is depends on the span of the rafters. For RAFTER spacing of 24" on CENTERS, 3/8" plywood is the minimum thickness where wood shingles or shakes, or asphalt shingles are used. For rafters 16" on centers, 5/16" plywood is the minimum. Thicker material can be used for smoother appearance and better RACKING resistance. If one is installing slate or some similarly heavy material, it is advisable to use 5/16" plywood as a minimum if rafters are spaced 24" and 1/2" if 16" OC.

Plywood should be installed in staggered pieces with the face grain perpendicular to the rafters. If trusses are used, it is unnecessary to stagger plywood end joints over alternate trusses, which simplifies decking layout and may reduce the number of cuts that have to be made.

Plywood decking should be fastened at each BEARING point, 6" on centers along the edges and 12" on centers along intermediate members. The number of nails used will depend on the amount of plywood used.

If the plywood is not of the exterior type, the edges should be at the gable end. The eave should be protected by the trim on the aluminum drip edge.

Most plywood decking that runs perpendicular to the roof framing must be supported by wood blocking or fastened together with metal fasteners called H-CLIPS. No special fastening is needed if the plywood has tongue-and-groove edges or is 1/8" thicker than the minimum thickness required by the spacing of the roof framing. *See illustration on following page.*

Structural flakeboard, another kind of roofing material, includes WAFERBOARD and OSB. They are installed like plywood: rafter or truss spacing, nailing, and edge treatments are the same, and the same thicknesses are used. Some structural flakeboard panels have flakes or strands aligned to increase the directional strength parallel to the length of the panel. These products should be laid with the long alignment dimension perpendicular to the supports.

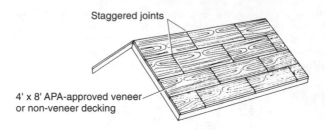

Staggered joints

4' x 8' APA-approved veneer
or non-veneer decking

Plywood roof deck

For years, boards were used as decking, but contractors changed to panelized material to save time and money. Boards can still be used and make a high quality decking, but installation is labor-intensive. Boards 6″– 8″ wide are used and should have a minimum thickness of ¾″ for normal rafter spacing (15″ and 24″). They should be nailed so that every rafter is supported by at least two boards. The boards are normally TONGUE-AND-GROOVE or SHIPLAPPED. In warm, humid climates where ventilation is very important, there may be spaces left between boards.

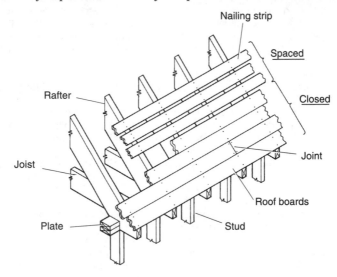

Nailing strip

Spaced

Rafter

Closed

Joist

Joint

Roof boards

Plate

Stud

Board decking

decorative panel Plywood panel with a "tooled" surface—rough-sawn, brushed, grooved, etc. Decorative panels are used both inside and outside the house for SIDING, PANELING, built-in, accent walls, counter facings, and displays.

deed restrictions These are laws that sometimes prohibit certain changes or additions to a property. For example, if a house is in an historic district, and is of historic interest itself, vinyl SIDING—because it gives a modern look—may be prohibited. Residents of a condo or planned unit development (PUD) should look for C. C. & Rs (covenants, codes and restrictions) before proceeding. One should also check any restrictions imposed on the property by a lender.

deflection Bending of a structural wood panel or framing member between supports under a load.

deformed bar Reinforcing bar with a rough or raised surface. This type of RE-BAR does a better job at reinforcing GROUT than a smooth rebar.

delamination When the plies in PLYWOOD come apart. When this occurs, the adhesive holding the plies together has failed. The most common reason for delamination is water, which breaks down the glue. Once this happens, the plies are not far behind.

demolition Disassembly, removal, and/or destruction of existing structures or materials.

> ### Unbuilding is an art, too
>
> Though not considered a building art, disassembling things properly certainly is. Doing it wrong can make the job take longer, be messier, costlier, and more hazardous. Doing it right makes the disassembly go smoothly and safely. For example, carrying a 500-pound cast iron tub out from a bathroom would be difficult, but breaking it into pieces with a sledgehammer first simplifies it enormously (see TEAROFF).

dewpoint Temperature at which point vapor in the air turns to water.

diaphragm Elements of a building that provide SHEAR strength to withstand WIND and EARTHQUAKE LOADS.

dimension lumber Any LUMBER over 2″ thick.

dimension shingles Roofing SHINGLES that are extra thick.

dimmer A type of switch used to control light emanating from a fixture.

dimples Slight depressions in Sheetrock where the nailheads are located. Dimples are a result of Sheetrock's being hammered in place; the Sheetrock is slightly depressed. These depressions are filled with a layer of JOINT COMPOUND as part of the finishing process. (This is required; a screw or nailheads showing in a finished plasterboard job automatically mars it.)

Hitting nails sometimes results in ripping the paper facing of the plasterboard. This break must be carefully patched with tape and joint compound. The facing paper is much less likely to be damaged if the plasterboard is secured with SCREWS.

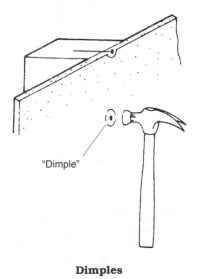

"Dimple"

Dimples

disposal tile field *See* ABSORPTION FIELD.

distribution box Used in a SEPTIC TANK system where there are two or more branch lines.

distribution panel Box within a house where the ELECTRIC SERVICE wires enter and are routed. The distribution panel is the heart of the electrical system. The service entrance conductors will be secured to heavy copper bars called BUSBARS. From there individual circuits will be routed off them.

distribution tile Pipe used in SEPTIC TANK system construction.

dog house dormer DORMER that looks like a small dog house.

dog's tooth BRICK laid with corners projecting from the face of a wall.

door Rectangular wood, metal, or fiberglass unit for weathertightness, security, and/or privacy.

Wood doors are generally broken into two broad categories: flush and sash, or panel, types. Both sides of the flush door are composed of flat, seamless panels. Sash or panel doors are composed of solid wood parts—framing and interior pieces. A sash door consists of STILES, vertical flanking members on the door; and rails, horizontal top and bottom pieces that are secured to the stiles. The stile on which the lock is housed is called the lock tile; while the stile where the hinges are is called the lock stile.

Sash doors may also have horizontal intermediate rails called cross rails. (The more rails there are, the stronger the door.)

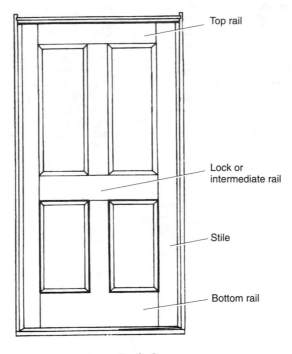

Sash door

Flush doors are also broken down into solid core and hollow core types. A hollow core door consists of a thin front and back panel and an interior hollow except for cardboard or other flimsy material. Hollow core doors, which have little or no strength, are designed for interior use. *See illustration on following page.*

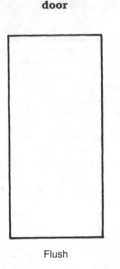

Flush

Flush door

Doors are also for use inside and on the exterior of the house, so-called "entry doors." The essential difference is the glue. Glue for exterior doors can withstand weather, while glue for interior doors cannot.

Doors come in a wide variety of sizes. Standard heights and widths increase in increments of 2", running from 1' to 3' 0" wide, and standard heights of 6', 6' 6", 6' 8" and 7' 0"; also 15" and 42" wide.

The most common interior and exterior doors used are:
- For bathrooms: 2' × 6' 8"
- For bedrooms: 2' 6" × 6' 8"
- Back door to house: 2' 6" or 2' 8" × 6' 8"
- Front door: 3' × 6' 8" (7 feet on older houses)

These are NOMINAL sizes but ACTUAL sizes of sash doors can be up to ¼" larger. Narrow doors are hinged together or used in pairs as bi-fold doors.

Hanging a door is not easy.(One carpenter with years of experience commented that the prospect still affects his stomach.) The main reason is that tolerances are close, and they have to be accurate. Too little space around a door can make it bind; too much, and it won't function properly.

The job starts with measuring for the door. If it is an entrance door, the chances are that the door will have to be planed on the edges to fit the existing framework properly. If an interior door is being installed, there is a 50-50 chance that the door will fit into the opening without any cutting. This relates to the way interior and exterior frameworks are installed.

In over 90 percent of the cases, conservatively speaking, a 6' 8" or a seven-foot-high door will be required. The door must be cut to fit the SADDLE so there is not more than ⅛" space all around. The existing door is removed, and then the WEATHERSTRIP, DOORSTOP, JAMB, and saddle are removed.

The carpenter selects a door as close to the opening size as possible to minimize the amount of trimming that needs to be done. The new door is stood inside the opening, and then the trimming process begins. The "horns," wood projections on top of the door to protect it, are cut off, and the door is placed in the opening.

From there, it is a process of simply trimming it to fit, using a saw and plane as required. Door openings often seem like parallelograms and trapezoids during this process. The job becomes increasingly laborious with the door put in place, excess marked with a pencil, removed, trimmed, and re-placed to see how it fits. This must be done many times.

When the door fits perfectly, it is mounted on hinges, the recesses for which are made with hammer and wood chisel. This, too, is an exacting process, and the carpenter must work very carefully to ensure that the door hangs straight and opens and closes properly. Many times it won't, and the carpenter will have to make adjustments in the placement of the hinges and/or the amount of material that is removed.

When the door fits perfectly and hangs well, it still must be trimmed on the bottom for any saddle. Installing the lock is the last step.

Finishing the door must be done exactly the way the manufacturer spec-ifies it. A number require that the bottom and top edges of the door be sealed to keep moisture out and prevent warping. If this is not done, the warranty against warpage can be voided.

Metal doors come in fewer styles than wood doors, but a number of man-ufacturers carry quite good-looking ones. Steel doors are made like flat metal boxes, and there are various ones to judge quality. Gauge of the steel will differ, being 22, 24, and 26. 26-gauge doors are not very good.

One way manufacturers like to advertise steel doors is with their R-VALUE. Some have cores filled with polystyrene or polyurethane. Other low-quality doors have no insulation at all. You may hear claims of doors with R-values as high as 15 (house walls are usually around 11), but this can be misleading. While the core material may be 15, the door may have LIGHTS in it which will allow a lot more heat out, dropping the average R-value of the door. Manufacturers usually have data sheets on the R-factors of their doors.

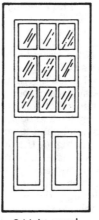

9-Light panel

Door with lights

Steel doors usually have excellent WEATHERSTRIPPING. Companies will say they're as tight as a "refrigerator door." Some steel doors come primed, and some come with jambs unpainted. When painting a door—any door—be particularly careful that you follow manufacturers' instructions. Doing it the wrong way can void the warranty.

Steel doors come PREHUNG, for the simple reason that you can't trim a steel door. But the ROUGH OPENING does not ordinarily have to be modified.

Fiberglass doors are designed to simulate fine wood. Like steel doors, they have a core of insulation and come prehung in a metal frame. Such doors are not cheap, costing about what a good oak door would cost.

See also DOOR BUCK, DOOR FRAME, DOOR LOCK, SLIDING DOOR *and* STORM DOOR.

door buck The ROUGH OPENING into which door framing is installed, or the framing on which the door is hung.

door frame The CASING into which a door closes. It consists of flanking vertical JAMB pieces and a horizontal section across the top called a head rail or LINTEL.

door lock Mechanism with or without knobs for opening and closing doors. Despite their name, some door locks do not have a locking function.

A door lock may be either interior or exterior. Exterior locks are built more ruggedly than interior ones, and manufacturers put more work into their design and finish. Door locks come in a wide variety or styles, and various finishes but are mostly BRASS-PLATED.

Cost of locks can vary widely from a few dollars to hundreds of dollars, but a higher price does not guarantee higher quality. The finish and style might be fancy and the working parts of the lock flimsy. On the other hand, a costly, plain lock with a brass-plated finish is likely to be good because the money has been put into the mechanism.

The key consideration when buying a lock is the "backset," the distance from the middle of the handle or knob to the edge of the door. Most locks have a $2\frac{3}{8}''$ backset and are installed in a $2\frac{1}{2}''$ hole.

Following is a lineup of exterior and interior lock types.
- **Interior:**
    - **Bathroom.** This looks like a "privacy" lock, but it is normally chrome-plated inside (brass-plated outside) to harmonize with other items normally chrome-covered in the bath.
    - **Passage.** This lock has no locking mechanism and is designed for use on doors where security is not a concern.
    - **Privacy.** This looks like the passage lock, but it has a button inside and a small hole outside into which a small screwdriver can be inserted to open the door in case of an emergency. This lock is useful when some degree of security is wanted, such as on a bedroom door. Pushing a button locks the door. Many of these locks come with a key or rod that enables them to be opened from the outside, a good feature if children lock themselves in a room.
- **Exterior:**
    - **Entry.** This lock has a knob inside and outside. It is opened with a key from the outside and a knob from the inside.

DEADLOCKS are also available for doors to provide an extra measure of security.

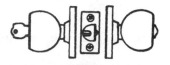

Entry lock

doorstop A device installed near the bottom of a door to ensure that it does not hit an adjacent wall when opened.

A solid doorstop is a slim, 3" brass-plated rod. One end has holes for mounting, and the other is fitted with a protective rubber pad. The doorstop may be mounted on the door, or it can be more permanently located on the baseboard if young children are present who like to stand on the doorstop and ride the door back and forth.

A spring-loaded doorstop is a tight, rigid spring with a plate for mounting. It works the same way as a solid type. It can be installed on door or wall. Because a spring-loaded doorstop bends when stepped on, it discourages free rides.

An adjustable doorstop looks like a spring with a threaded end. When installed on a door hinge, it stops the door from banging into a wall when it is not possible to install a stop in a regular position.

dormer A roofed projection from a sloping roof into which a window is set.

Dormer derives from the word dormitory, or sleeping place.

The dormitory, located in the attic, was lit by windows which came to be known as dormers. *See also* DOG HOUSE DORMER.

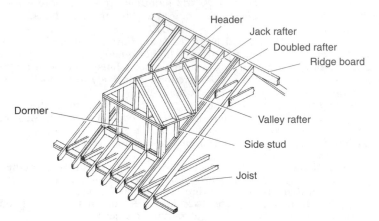

Dormer: dormer framing

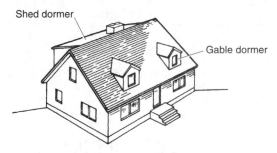

Dormer: shed dormer and gable dormer

double header In building, a pair of headers instead of one.

double-hung window WINDOW with upper and lower sashes that slide up and down to close and open. *See also* WINDOW. *See illustration on following page.*

double pitch skylight SKYLIGHT that slopes in two directions.

double rafter Pair of RAFTERS installed side-by-side.

doug fir Short for Douglas fir, the most common type of wood used in the making of PLYWOOD.

doughnut The plastic or wax ring that the HORN of a toilet is inserted in to prevent leaks. Wax rings are pushed against the top of the toilet waste pipe (CLOSET BEND), and the horn or bottom of the toilet bowl is pressed into

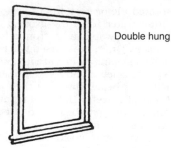

Double hung

Double-hung window

this, making a watertight seal. Wax rings come single- and double-sized, and even with a plastic sleeve. The wax ring with a plastic sleeve ensures a leak-free seal and can help if the bowl is a little lopsided; the ring with the sleeve will still do the job.

dovetail A type of joint used in woodworking, the cuts resembling the tail of a dove. Framing members do not utilize dovetail joints, but they are quite common in the making of drawers.

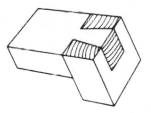

Dovetail; dovetail joint

dowel 1. A short, round wooden stick with ends cut flat. Dowels have many uses, but they are chiefly used to re-enforce the corners of tables and other furniture and KITCHEN CABINETS. They come in a wide variety of sizes. **2.** Straight metal bars used to connect sections of masonry.

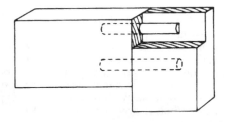

Doweled joint

downspout The vertical portion of a gutter system. Also known as a leader, the downspout carries water away from the house. Otherwise, water could collect around the foundation and find its way into the basement, the number one cause of leaks in basements.

drain A general term describing a variety of pipes or stationary drains in tub and the like, which carry water away. *See also* DWV SYSTEM.

drain (drainage) tile 1. Perforated tile or sections of perforated plastic or bituminous pipe used to drain water from the ground around the house. Drain tiles are valuable in routing water away from a foundation, where it could flow in and cause damage. Drain tile is used where water is perceived as a possible problem, such as sloping or low areas; foundations enclosing basements; or habitable spaces below the outside finish grade.

The drains are installed at or below the level they are to protect. Normally, they are set on a 2″ bed of gravel with 6–8″ of gravel above, but some soils are very permeable and the tiles can be placed directly on them. Openings at joints are sealed with asphalt strips.

The water that is routed away may go to an outfall or ditch; in some communities a DRY WELL may be used.

2. Pipe—also usually bituminous—used to drain off effluent in a SEPTIC TANK system.

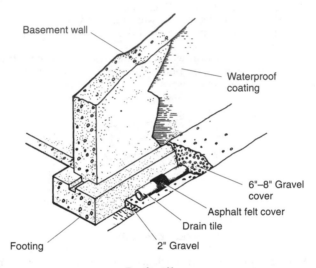

Basement wall

Waterproof coating

6"–8" Gravel cover

Asphalt felt cover

Drain tile

Footing

2" Gravel

Drain tile

dressed The final size of a piece of wood. BOARDS and other pieces of lumber go through a number of finishing operations before getting to their final, or dressed size.

drip 1. Structural member of a CORNICE or other horizontal exterior finish course that has a projection beyond the parts for water run-off. Also called a "drip cap." *See illustration on following page.* **2.** A groove in the underside of a SILL or drip cap to cause water to run off on the outer edge.

drip edge Overlapping piece of metal around the roof. A drip edge keeps water from running down the FASCIA and getting under the roof.

dripstone Stone section over a window to prevent water from running down it.

drop The bottom end of an open NEWEL post. Some newel posts terminate at the floor, but others do not. A decorative part called a drop is secured to provide good looks.

drop ell A plumbing FITTING shaped like the letter L.

drops Short for drop cloths, covers used to protect areas from paint spatters. Standard drop cloths are canvas and come in a variety of sizes, the most common being 9′ by 12′. Painters carry a variety from job to job.

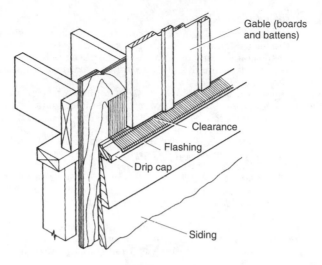

Gable (boards and battens)

Clearance

Flashing

Drip cap

Siding

Drip cap

drop siding *See* SIDING.

drop tee Plumbing FITTING.

dry mix Packages of dry mix that contain all the ingredients necessary for making CONCRETE or MORTAR by simply adding water.

dry rot Describes the brown, crumbly appearance wood gets when it has been attacked by fungi. "Dry rot" is actually a misnomer because fungi can only function in situations where wood is damp.

Fungi do not necessarily need damp wood to begin with. Some kinds of fungi can carry water—mostly from the soil—setting up a situation conducive to dry rot. Rotted areas can eventually dry.

There are a number of products on the market that can save wood that has dry rot. With one, where holes are drilled into the wood, and a liquid hardener is injected that effectively turns the wood into plastic. Another product involves scraping the dry rot away and applying a putty-like filler that dries hard.

drywall Another name for plasterboard, wallboard, or Sheetrock. Drywall comes in a variety of sizes: 4' wide in 6', 7', 8', 10', 12', and 16' lengths and ⅜", ½", and ⅝" thicknesses. The ⅝" thickness is used mainly to fireproof walls and comes only in the 4' × 8' size because of its weight. As building materials go, drywall, also known as gypsum board, plasterboard, and most commonly the brand name Sheetrock, is uniform in quality, regardless of manufacturer. There are no hidden defects; it is visibly good or broken. And it is cheap.

Plasterboard is primarily used as a new wall material. It is also good for covering up heavily cracked walls and ceilings. If you do not like the look of ceiling 0tile, plasterboard will cover it nicely.

Prior to the introduction of drywall in the 1900s, interior walls were mostly made with PLASTER, which was known as Wetwall. This still requires a high degree of skill and plenty of time to install. Drywall is, in effect, ready-made walls. Its installation does not require nearly the skill plaster does. The material has revolutionized the building industry.

Hanging drywall

When professionals "ROCK" a new room, they will first consider how to do it with the fewest joints. This minimizes the taping that must be done to seal seams or joints. This usually means the boards are installed horizontally because it is much easier to tape a seam at shoulder height than from floor to ceiling.

Ceiling boards are installed first. One is lifted up in a corner, held against the ceiling, and nailed in place. If possible, installers will try to get drywall long enough to span the ceiling in one piece. If not, different-length panels are cut with a utility knife so that the ends will not all land on one joist. DRYWALL NAILS are used, with hammer blows DIMPLING the surface. These dimples are filled later. The panels have tapered edges on the long edges. These are butted against one another.

SCREWS are also used. These are said to be better than nails. For one thing, they're faster because they are screwed in place with an electric screwgun and leave shallower dimples than nails. And, as with screws on any job, the boards are more secure.

At some point—usually the second panel in the average house—a drywall panel will have to be cut to fit around a ceiling fixture. This is done by measuring where the hole will fall on the panel and cutting it out prior to installing the board.

Rockers will also try to pick the longest practical boards for walls, and work things out so there are as few joints as possible. In general, the rockers work from the top down, starting at any end. A panel is hiked tightly up in a corner, then tacked in place and nailed to STUDS, which have been previously marked off on the ceiling and PLATE.

The bottom board is hiked in place under the first and nailed in. This procedure is followed all around the room with the tapered edges of the boards meeting.

When all boards are in place, taping is done.

The vertical corners are taped. Joints between ceiling and walls are optional; professionals use tape, cover molding, or crown molding. Cover molding is preferred because it involves less matching at corners.

The tape itself is 2" wide with holes that are relatively large or almost invisible. The latter is preferred because the joint compound used does not ooze through the holes. There is a new type of paper available with its surface mechanically roughened.

Joint compound, commonly called mud in the trade, comes in standard and lightweight versions. It is applied—BUTTERED—in conjunction with tape in a series of coats. Using a broad knife, the installer applies a thin coat of it about 6" wide and feather the compound smooth until its edges "disappear" into the drywall—over the joint. The tape will be embedded in this and the board knife used to draw it flat and flush, applying a thin coat over the tape. This is allowed to dry, and another coat, about 12" wide is applied, feathered out, and allowed to dry. A final or "polish" coat is applied over this. Corners are handled by applying vertically folded strips of tape embedded and covered with compound. This is done with a regular joint knife or an angle knife known in the trade as an "angle plow." Professionals also use a device called a bazooka that applies tape and compound at the same time.

Outside corners have to be protected against impacts; tape and compound are not enough. Corner bead is used for this. This is a perforated V-shaped metal strip. It is slipped onto the corner and then nailed, stapled, or crimped in place. Additional compound is applied and feathered over the outside corners.

Sanding is not required on most drywall jobs. If it is, the wall should be primed first to reduce the chance of roughening the paper surface of the drywall. Some installers use wet sponges as a final finishing tool to ensure smoothness. Some rockers apply a "skim coat," a thin coat of mud over the entire wall to create an absolutely uniform surface.

When sales went off the wall

The core of drywall is gypsum, a naturally occurring rock of virtually bottomless stores. Originally used to make plaster, gypsum was adapted by Augustine Sackett for use in making the first drywall. Sackett made a sandwich of plasterboard composed of four layers of tough paper sandwiched between three layers of plaster of paris and called it "Sackett Plaster Board." While it was as heavy as it was weak, builders found it useful. Plastering was a three-coat job, but the builder could nail the plasterboard to the studs and cover it with a relatively thin coat of plaster—and that was the finished wall. As one can see, this cut the time required for finishing jobs greatly.

In 1909, Sackett sold his burgeoning company to US Gypsum, which redesigned the product, eliminating the inner layers of paper and improving it in other ways. In 1917, they introduced this new product as Sheetrock, and it was here to stay. Today billions of square feet of the product are used yearly worldwide.

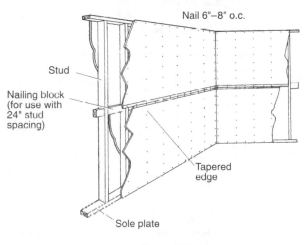

Drywall

dry well A rock- and gravel-lined hole in the ground designed to receive waste water from the house but not sewage. *See illustration on following page.*

ducts Round or rectangular pipes for circulating warm air in a forced-air heating or air-conditioning system. *See also* BOOT.

dusting Powdery material on newly placed CONCRETE.

dutchman A piece designed to fill a gap when repainting something.

The 17th century was a time when Dutch-English relations were strained, and a number of slurs arose for each nationality. "Dutchman" came to mean a cheap way of doing something instead of a proper, more costly one.

The term is commonly used today to describe a section of wallpaper used to fill a gap or a filler piece of wood in carpentry, such as when a table leg is too short. The term can also apply to stonework. The September 3, 1960 issue of the *New Yorker* magazine stated: "He mended the (marble) lion by cutting recesses several inches deep wherever the stone was damaged, and fitting new pieces of stone therein. These pieces are known in the trade as dutchmen."

dwarf wall A short wall that does not go all the way to the ceiling.

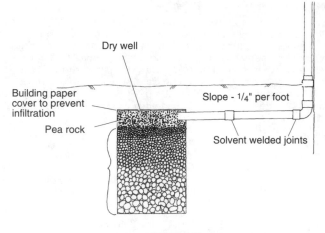

Dry well

Dry well

DWV system Stands for drain waste-vent, and describes a home's complete system for the removal of waste. Unlike the WATER SUPPLY SYSTEM, which depends on water pressure to work, the drainage system depends on gravity.

The DWV system consists of the drain, waste, and vent systems. Each works independently of the others. The drain system takes away water, the waste system excreta, and the vent system circulates air and provides air pressure throughout the system.

In a typical home plumbing system, waste water drains from a sink, and goes through a series of 1″– 2″ diameter pipes that are sloped and lead to the soil stack, a 3″ or 4″ diameter pipe. The stack—and there is usually just one in the average house—runs vertically from the basement or lowest floor. When a toilet flushes, the waste is carried by a short soil pipe called a "closet bend" directly to the stack. All the waster water and excreta runs down the stack to another large diameter pipe, called the "building drain," which slopes horizontally to the sewer line, the name given to the building drain after it exits the house. The sewer line carries everything to a CESSPOOL, SEPTIC TANK, or public sewer.

At various points in the system are TRAPS. These are safety devices, loops of pipe under every fixture in the house and built into the toilets. They are designed to keep a permanent seal of water against sewer gas or vermin entering the house.

A plumbing system also contains vents, which bleed off noxious gases and, importantly, allow atmospheric pressure on trap seals to keep them in place. In a house, the vent may only be the upper portion of the stack, the point directly above which the highest fixture drains. In other cases, there may be pipes with a vent-only function.

"Cleanouts," also known as cleanout plugs, are threaded metal or plastic caps screwed into pipes, usually at the bottom of the stack and at points where drain pipes change direction. These points are where clogs often occur. The plugs can be removed and a drain-clearing tool inserted. *See illustration on following page.*

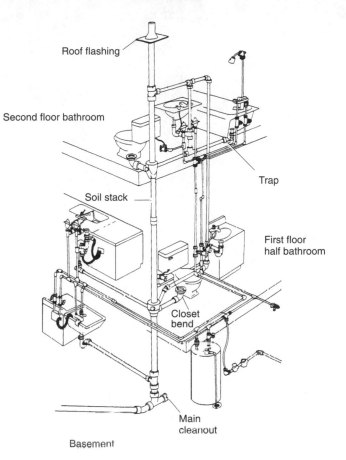

Roof flashing

Second floor bathroom

Trap

Soil stack

First floor
half bathroom

Closet
bend

Main
cleanout

Basement

DWV system

ears **1.** On KITCHEN CABINETS, projecting pieces of wood on the sides that allow the cabinet to be trimmed to fill a space totally. **2.** Projecting metal ear-like tabs on electrical boxes that allow installion of the BOX without it falling through the wall opening. Also known as plaster ears. If necessary, they can be clipped off. *See illustration on following page.*

earthquake load Also known as "seismic" load, this is a measurement of how much stress a building can take during an earthquake. Earthquakes only occur in certain areas of the US, of course, and many areas need not be concerned about them. Some, of course, do.

An earthquake is an adjustment masses of earth make. Stresses are put on the earth, and to relieve them, the masses move along what seismologists call a fault line. It is not farfetched to compare this to what happens to a plaster wall as the house settles—the stresses produce a crack.

Such movement moves the earth sharply on the horizontal plane, jolting many buildings on it according to the severity of the quake. It has been found that wood-frame houses are able to take brief jolts of great magnitude without being damaged. It is estimated that houses built to take WIND LOADS of fifteen pounds per square foot can resist most earthquakes.

MASONRY structures are a different story. Unlike wood-frame construction, masonry is not resilient. An irresistible force meets the immovable object and damage occurs. In areas where earthquakes are prevalent, masonry structures must be designed to withstand the shock. *See also* DEAD LOAD, *and* LIVE LOAD.

easement rights These laws concern the right to use land owned by someone else. Utility companies commonly have easement rights because their power lines cross someone else's property.

eave The lower margin of a roof projecting over a wall.

87

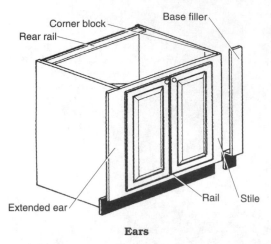

Ears

Eave derives from the Old English word "off," which means over. Another name for eaves is CORNICE, of which there are a variety of types.

eave vent An opening in a roof or eave to allow for passage of air.

Houses that are tightly constructed and insulated generate warm air that can create CONDENSATION, doing damage to a building's interior. Vents and many times fans are used to bleed off this warm air. *See also* ATTIC VENTILATORS, GABLE VENTS, *and* SOFFIT VENTS.

edgenailing Nailing into the edge of a BOARD.

edge treatment How PLYWOOD edges are finished, such as covering them with wood strips or wood banding, or filling them with putty or Spackle. If the edges of the plywood are kept square and a strip is applied, the process is "butt edging." "Mitered butt edging" involves mitering the edge and securing a wood strip.

Edge treatment

edge void PLYWOOD panel defect where there is a gap in the edge of a panel. Such gaps involve inner plies splitting or breaking away during manufacture.

efflorescence White, powdery deposits on masonry. These deposits are actually soluble salts in the material that have been dissolved by water and migrated to the surface. Efflorescence cannot be successfully painted over. The powder must be brushed off and the residue cleaned off with a dilute solution of muriatic acid.

egg-and-dart molding MOLDING with a tooled finish consisting of alternating egg and dart shapes.

elbow A type of plumbing FITTING that enables water to flow in a curve. This fitting is roughly shaped like an elbow, and is available in 90 to 45 degrees.

90° Sanitary elbow

Elbow

electric service Used to describe the electric power supplied by a utility.
Electric service to a house may be any of three capacities.
- Two-wire 115 VOLT service, where one of the conductors is connected to the various electrical devices.
- Two-wire 230 VOLT service is when both wires are connected to the electrical item.
- Three-wire, 115/230 volt service.

Electric service may be provided above the ground (*see* OVERHEAD SERVICE) or buried in it (*see* UNDERGROUND SERVICE). *See illustration on following page.*

embedment 1. In general, the placing of one material into another so the two become one. **2.** Placing a FELT, AGGREGATE, mats, or panel into hot BITUMEN.

embossed 1. Plywood panel surface treatment that leaves textured design in the face but leaves the panel paintable. **2.** In general, design impressed into a surface.

emulsion Product that contains oil and water and does not separate.

enamel A glossy paint.
Years ago, enamel was short for enamel PAINT, a hard, shiny oil-based coating. As oil paint faded and latex paints came to the fore, the term was used to describe all kinds of paint, including "flat enamels," a sort of contradiction in terms.
Years ago, oil-based paint was hard and durable, so the meaning has stayed with people: any time they hear the term "enamel," they assume it means "durable." It may or may not, depending on the paint. Another meaning that has endured is "shiny."
Glossy paint has certain advantages over flatter paints—and some disadvantages. Glossy paint is less likely to absorb dirt than flat paint, and is therefore easier to keep clean. The surface of glossy paint is tighter and denser, a more difficult surface for mildew to take hold on. Some manufacturers guarantee a paint as mildewproof simply for being glossy.
On the negative side, glossy paints are thinner than flat paints and tend to show up mars and dings in surface more readily. Application is a little more difficult; brush marks show more readily with a shiny paint than with a flat one.

enameled brick Brick with a shiny surface.

encumbrances RESTRICTIVE COVENANTS that run with a particular piece of land and can affect its marketability and value. Such covenants may limit the density of buildings per acre; regulate size, style, or price range of the building to be erected; or prevent particular businesses from operating in a particular area.

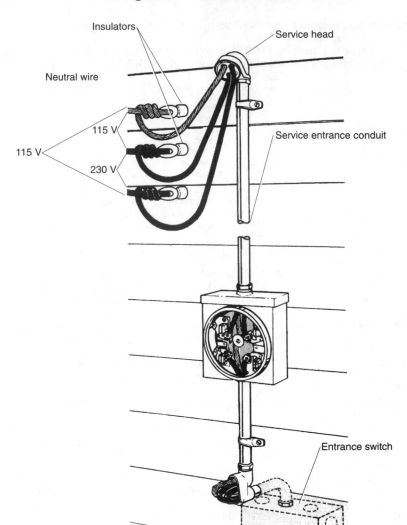

Insulators

Service head

Neutral wire

115 V

115 V

230 V

Service entrance conduit

Entrance switch

Electric service

end grain The end of a piece of wood when wood is cut across the grain (*see* WOOD GRAIN). All structural wood panels are end grain and must be finished.
    Wood consists of a series of longitudinal tubes. The ends of these tubes are open on the end grain of a board. If the wood is not given an impervious coating, water can get in and do damage. *See illustration on following page.*

endnailing Nailing into the end of a board.

engineered 24″ framing Method of framing a building using 24″ rather than standard 16″ spacing. *See* O.C.

envelope Continuous FELT fold made by wrapping and securing a portion of a base felt back up and over the felt plies above it.

environmental codes Laws governing building that can have an impact on the environment. *See also* BUILDING CODE.

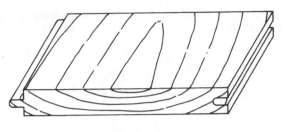

End grain

EPS forms Expanded polystyrene, forms that encase CONCRETE, staying in place and acting as insulation.

Foam is the latest thing in forms for FOUNDATIONS. Foam comes in panels and block. The panels are basically CAVITY WALLS reinforced with steel REBAR. The panels can be linked together with plastic ties. The panels are light and can be cut with a handsaw. When set in place, the panels are filled with concrete. Siding can be attached to clips on the panels.

There are three basic foam block systems. In one system, there are lightweight BLOCKS with foam inserts, containing polystyrene pellets in the AGGREGATE. The blocks are dry-stacked and covered with a structural finish coating. Another system uses dry-stacked block that accept 2″-thick foam inserts. The foundation is coated with a fiberglass-reinforced surfacing. The third system uses pre-insulated BLOCK the CORES of which have been filed with foam and which partially cover the WEBS on the blocks, taking the place of and tending to be of more insulation value than mere MORTAR.

Foam block systems have been largely confined to foundation walls. Though more expensive than conventional methods, contractors say that the foundations are easier to install (for one thing they are light), requiring fewer labor hours, and the insulation value of the foam gives the average consumer a big payback. "Above grade"—above ground level—walls have R-values of from R-15 to R-25.

EPS forms also allow for pouring concrete in sub-zero weather. Manufacturers say that the foam-encased concrete can stand up well in hurricane country. EPS forms have been used in Europe for twenty years, but not as long in the States, where they have about one percent of the market at this writing. Whether or not EPS forms will take hold remains to be seen. Some contractors say the product is still too new to evaluate properly. That will take time.

On the negative side, some of the older EPS forms were subject to blowing out. (Manufacturers say the foam was not dense enough.) And while insects were not drawn to the foam itself, they could burrow through it to the wood members. And there is the matter of simple perception: home buyers may wonder just how good foam can be.

escutcheon 1. Metal shield around keyhole to protect wood. **2.** Fancy metal plate to which a doorknocker is secured. **3.** In plumbing, the decorative circle of metal around wall faucets.

European cabinets European, or frameless, cabinets are made like FACE-FRAMED CABINETS except the doors are flush.

European cabinets, which originated in Germany in the early thirties, are built as a box like their face-frame counterparts. The front raw edges are covered with thin strips of laminate. The flush doors are completely covered, and the doors are hinged on their backs and inside the cabinet. No HINGES are visible by looking at the cabinet, and the front is seamless except for the juncture where the doors meet. Many people feel that European cabinets have a slick, modern look that can't be topped; but many others favor the more traditional look of face-framed cabinets.

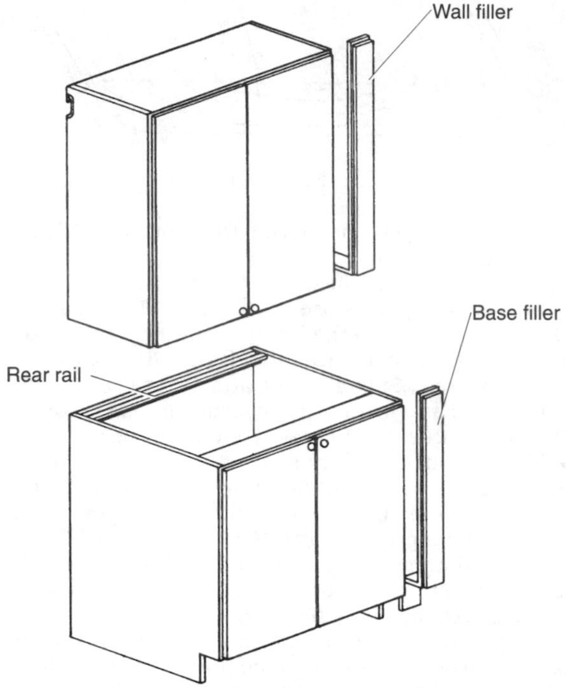

Wall filler

Base filler

Rear rail

Base cabinet

European cabinets

excavation The removal of earth at a building site.

Excavation follows STAKING OUT the perimeter of the building. First, a bulldozer and front end loader remove the top soil and stockpile it for later use. Excavation goes down to a point where the tops of the FOOTINGS will be, or the bottom of the basement floor because some soils become soft upon exposure to air or water. Unless formboards are to be used, it is not advisable to make the final excavation for footings until it is time to pour the concrete.

Excavation is done wider than the limits of the building to allow room to work in when constructing, waterproofing the foundation wall, and laying DRAIN TILE, if necessary. Just how steep the walls of the hole will be depends on the soil. If the excavation is in solid clay, they can be almost vertical; if sandy, the walls must of necessity be sloped.

Some contractors only rough-stake the perimeter of the building for removal of soil. When the proper floor elevation has been reached, the footing layout is made and the soil is removed to form the footing. After the concrete for the footings is poured and set, the foundation is established on the footings and marked for the placement of the formwork of concrete BLOCK wall.

excavation line Line made on a building site to mark off the area to be excavated for a building. *See also* EXCAVATION.

exhaust fan Fan that draws air from any given area.

Ventilating a bathroom is important not only to remove odors, but, as much as possible, to remove moisture. Unchecked, moisture can severely damage paint, wallpaper, and so on.

The average bathroom generates a lot of water vapor. Because bathrooms are ordinarily warm, much of this vapor is retained. Exhaust fans help to remove this moisture-laden air. Some exhaust fans have just an exhaust function, but others have built-in lights and heaters.

The key consideration for an exhaust fan is that it does not remove the air too quickly. Doing so can give rise to a chilly environment. The idea is to remove the air so the water vapor cannot do damage but not so fast that the room becomes uncomfortable.

Ventilation experts suggest using a fan capable of changing the air in a room eight times an hour. Air exchange is measured in cubic feet per minute or CFM. This will be marked on the fan. The formula for getting the CFM for a particular room is as follows: multiply the area of the room by 1.7 and round it off. For example, a 5′ by 7′ bathroom would be 35 square feet. Multiplying that by 1.7 gets 59.7. Rounded off, that is a 60 CFM fan. (Note that the calculation assumes an 8′ ceiling.)

Noise is another consideration vis-a-vis exhaust fans. Some have centrifugal rather than impeller blades; some are insulated; some are mounted on rubber; and some are big enough to resist vibration while some are not. These factors all contribute to the noise the fan makes when running, and this is measured in "sones." This is measured on a scale of 1 to 10. Experts recommend that fans not exceed three sones, something that will be imprinted on the fan. *See also* CONDENSATION.

expansion joint 1. Bituminous fiber strips used to separate BLOCKS or units of CONCRETE to allow the joint to crack rather than the material if the material moves because of temperature changes. **2.** Open vertical space in block walls to allow for movement in case of temperature changes.

expletive A stone used to fill a cavity in masonry.

exposed joint Any mortar joint on the face of MASONRY that is above ground level.

exterior Refers to the outside of a building.

Work is generally divided in builders' minds between EXTERIOR and interior work. Generally, work outside the home is more physically taxing.

facade The front of a building. Facade is normally used to describe buildings bigger or more elegant than a home.

face 1. In general, the exposed portion of a structure. 2. The high-grade exposed side of a panel. For example if a PLYWOOD panel is A-B, the face would be A.

face brick *See* BRICK.

faced wall In masonry, a COMPOSITE WALL in which the MASONRY facing and backing react the same way when under load.

face-framed cabinets Cabinets with a separate framework in the front.

Face-framed cabinets have been around a long time. Essentially, they are a box without a front. Face-framed cabinets may be made of PLYWOOD, MDF, PARTICLEBOARD, or other dimensionally stable materials (individual boards are usually not used). The cabinet front is made of solid boards, usually joined with dowel joints. It may be secured to the front of an individual cabinet or over a group of them. Either way, it gives the cabinets a recessed look and a place to secure door hinges to.

There are three kinds of doors used on face-framed cabinets: "flush," "overlay," and "lipped" or recessed. The flush door fits within the cabinet frame, its face flush with the cabinet framework. The overlay door slightly overlaps the cabinet opening all around. The lipped door has a rabbet, or recess, cut in the edge.

Cabinet makers in America like face-framed cabinets because the framework helps make the cabinet stronger, while allowing adjustments to correct the cabinet size and alignment. *See also* EUROPEAN CABINETS. *See illustrations on following page.*

face grain Direction of the grain of the face of a veneer-faced panel in relation to its supports. A panel's greatest stiffness and strength is parallel to the face grain; therefore it is common practice to run the face grain or long

95

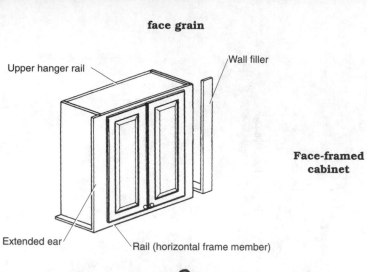

Upper hanger rail

Wall filler

Face-framed cabinet

Extended ear

Rail (horizontal frame member)

Flush

Overlay

Face-framed cabinet doors

Lipped

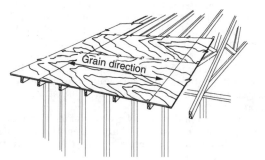

Grain direction

Face grain

dimension of the panel across supports for the greatest stiffness and strength. *See also* PLYWOOD. *See illustration on previous page.*

face line The lines masons follow when building walls.

Face lines must be accurate because the straightness of the walls depend on their being so. As such, strong string is stretched out and secured to boards staked in the ground. *See also* CROWDING THE LINE *and* MORTAR JOINTS.

facenailed Nails that have been driven in perpendicular to the initial surface being penetrated. Also known as direct nailing.

face shell The front of a concrete block. *See* BLOCK.

false header 1. A short length of lumber fitted between two joists. **2.** In masonry, a half BRICK used in the Flemish BOND pattern.

false rafter Extension to a rafter over a CORNICE, particularly where the roof line has been changed.

fanlight Any window over a door.

Technically, a fanlight is a window shaped like a fan, but the term has come to mean any window over a door. The original fanlights were commonly installed over Georgian entry doors; the hall could be illuminated without privacy being compromised.

fascia Horizontal trim pieces that run around a house at eave/wall junctures.

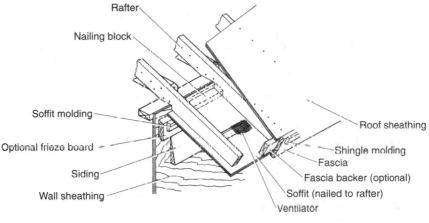

Rafter

Nailing block

Soffit molding

Optional frieze board

Siding

Wall sheathing

Roof sheathing

Shingle molding

Fascia

Fascia backer (optional)

Soffit (nailed to rafter)

Ventilator

Fascia board

fascia backer The main structural support member to which the fascia is nailed.

fastener General term for hardware used to connect building materials. Fasteners are chiefly nails and screws, but a variety of other items are also considered fasteners. *See* FRAMING FASTENER.

fat lime Quicklime made by burning a pure or nearly pure limestone. Sometimes called "rich" lime, fat lime is used in PLASTER jobs.

fat mortar MORTAR that is rich in lime.

faucet Mechanism for controlling water flow. Faucets are known as "valves" in the trade, because they control water flow.

The key consideration in selecting a faucet is the distance from the center of the hot water faucet handle to the cold one. This distance is called the CENTERS. Tappings are made in sinks and lavatories to certain centers. Hot and cold water supply pipes will have the same centers. Lavatory faucets usually have 4″ centers, but eight-inch centers are not uncommon. Wall-mounted lavatory faucets usually have 4½″ centers, but may be 6″.

Kitchen sink faucets are usually 8″ apart, and so are tub faucets. Utility sink faucets are normally 4″ apart.

Faucets may be of the compression or washerless type. In the compression type, the handle turns a stem. On the bottom of a stem is a washer that presses against a "seat." Turning the faucet handle to off presses the washer against the seat, stopping water flow. Release it, and the water flows.

Washerless faucets have working parts that do not, in a strict sense, use washers like a compression faucet does (*see* FAUCET). Washerless faucets come in two basic styles—ball and cartridge—both working essentially the same way. On the ball type, the water emerges from either of two holes in the faucet body; the holes contain springs and washer-like pieces. When in the off position, the ball's solid parts block the water ports. To turn it on, the ball is manipulated so that holes in the ball align with the water ports and the water flows through them and out the spout. The same happens with the cartridge faucet, a one-piece cylinder with holes on one end. When these holes are aligned with the water ports, the water flows. Both cartridge and ball types have a single handle.

Compression faucets utilize rubber, plastic, or, in newer models, ceramic discs. The discs last much longer, but some plumbers say that unless the plumbing is new they may not be a good investment: scales from old pipes score the disks, rendering them ineffective.

Faucets are made from a variety of materials, including plastic, chrome-plated plastic, chrome-plated pot metal, chrome-plated brass, and plain plastic. This last is referred to in the trade as "builder's special."

One thing that is often forgotten is the ease of operating a faucet: it should be possible to turn a faucet even with wet and soapy hands. Faucets with elaborate shapes may be good to look at, but they can be inconvenient when you are trying to turn them off and on.

feathered Soft, plastic material worked so it gradually blends into surrounding flat surface.

Experience, the only teacher

PLASTER and JOINT COMPOUND are two materials for which experience is the only teacher in applying them properly. Of the two, plastering is the more difficult, but neither is simple for the inexperienced. The reason is that one must have a certain "touch," and this takes a long time to develop.

felt papers A fabric made of glass fibers (glass-fiber felts), vegetable fibers (organic felts), or asbestos fibers (asbestos felts).

female In plumbing, the FITTING or pipe that receives the male end of a pipe or fitting.

fender wall A low brick wall around the fireplace hearthstone.

fenestration The way windows are arranged in a particular building.

fiberboard General term to describe sheet products made with wood fibers. Fiberboard is essentially made of scraps of wood and wood fibers. They are mixed together with glue and pressed out. The resulting product is strong, though not as strong as pure wood.

Fiberboard is much cheaper than wood or plywood, making it a favored product for many contractors who use it well in various situations.

See also HARDBOARD, MDF, OSB, PARTICLEBOARD, *and* WAFERBOARD.

fieldstone Stone found in nature and used more or less as is.

field tile Tile in the middle of a wall, as opposed to the edges.

figure Wavy, mottled, or streaked look of wood GRAIN.

filler block Piece of 2 × 4 installed between a three STUD corner POST. Filler blocks add rigidity to corner post, and provide more nailing surface. *See* FRAME.

fillet Strip of wood or line of masonry with a CONCAVE face connecting two surfaces meeting at an angle. Such a strip adds a measure of good looks it might otherwise not have.

fingerjoint Joints in wood that interlock like interlocking fingers.

finger room Narrow space left behind a BRICK VENEER WALL and house sheathing so that the mason can get his fingers in to facilitate installation of the veneer.

　　Virtually all brick walls are not structural components; they are just neatly stacked up, mortared together, and tied to the house sheathing with metal ties. For the mason to make these ties, he needs to have a little room behind the brick.

finial Decorative piece at the top of something such as a NEWEL post, chair, or roof peak.

finish coat The final coat of a variety of different materials.

　　Many construction facets involve multiple coats of material. PAINT, STUCCO, and PLASTER are just a few. The finish coat is the last one; it is where the craftsman make the work as good-looking and mechanically correct as possible.

finish grade The level of the land following construction. *See also* GRADE.

finishing Applying the final coat and imparting the final look to a surface.

firebox The metal enclosure in a fireplace where the fire is confined. *See* CHIMNEY.

firebrick BRICK in a fireplace that fire contacts and that is made to withstand high temperatures without cracking.

fire clay Clay that can withstand high heat. Fire clay is used in the making of FIREBRICK.

fireplace Lower part of a CHIMNEY, consisting of the FIREBOX, throat, damper, SMOKE CHAMBER, and shelf.

fire-rated system Roof, floor, and walls that have been tested for fire safety. Testing for such things as flame spread rate and fire resistance is performed by agencies such as UNDERWRITER'S LABORATORIES. A one-hour rating, for example, means that an assembly similar to the one tested will neither collapse nor transmit flame or high temperature for at least one hour after a fire starts. *See also* FIRESTOPS.

firestop A solid, tight closure of a concealed space, placed to prevent the spread of fire and smoke.

　　In a stick-framed wall, firestops are horizontal pieces of wood called cats. Masonry is installed between joists of a framed wall to prevent the spread of fire.

fish The process of running electrical wires through walls. This is often needed on rehab work. Cable is taped to a long thin tool called a "fish tape" and fed through drilled holes in framing members.

fisheye A problem with clear finishes wherein the finish takes on the appearance of the eye of fish with distorted circles.

fittings In plumbing, fittings perform three jobs: they continue straight runs of pipe; they enable pipe to make turns; and they allow changing from one type of pipe to another, such as galvanized to plastic.

　　Fittings for waste and water come in essentially the same shapes, though drain or DWV fittings are bigger. Following is a roundup of common types:

　　• **Compression.** A short, threaded sleeve with a compression ring and nut. Compression fittings are only used for joining water pipe. Plumbers use them where SOLDERING—using an open flame— might be difficult or where the heat might damage a component, such as a valve.

　　• **CPVC.** Fittings designed to be used with plastic hot water pipe.

- **Drainage.** These look like standardized galvanized fittings, but the walls are much thicker. Used in the same way as standardized galvanized fittings, the inside of a drainage fitting forms a smoother surface, greatly minimizing the chance of blockages.
- **Flare.** Externally, this looks like compression fitting. The main difference is that the ends have been flared—made horn-like—with a flaring tool. Flares, as they are called, are more secure than compression fittings. For this reason they are most often used for lines that carry oil, gas, and other flammable liquids.
- **Hubless cast iron.** Cast iron pipe with no formed ends; joinery with no-hub connectors. For years, cast iron PIPE was joined with hot lead, but now connections can be made with the hubless type using no-hub connectors. Joinery is simple. Each of the plain ends slip into a neoprene gasket and is secured with stainless steel clamps. The clamps are tightened with a wrench that slips at 48 pounds, making overtightening impossible. No-hub cast iron pipe is not normally used in the ground. Wherever it is, care must be taken to support the joints so they do not flex. *See also* WIPED JOINT.
- **PVC.** Plastic fitting (polyvinyl chloride) used on cold water pipe or drainpipe. Fittings made from PVC are assembled, like other plastic fittings, with a solvent that actually melts the plastic and welds the fitting and pipe end together. It sets quickly, so the plumber assembles the pipe dry and marks where they align and the depth that the pipe is to penetrate the fitting.
- **Standard galvanized.** These are short, variously shaped sections of silvery pipe with internally threaded ends. These fittings are about ⅛″ thick. They are always FEMALE; MALE ends of pipe are screwed into them. They are used for both water and waste pipe.
- **Sweat.** A smooth copper fitting in various shapes. Sweat fittings also come made of cast brass, which is lighter colored than copper or wrought (extruded) copper, and are used for both water and waste. A more accurate name for the fitting might be "solder fitting" because hot solder is used to join the fitting to the pipe. This method of joinery is used for both water and DWV pipe. Brass is used wherever water is corrosive. *See also* SWEATING.
- **Transition.** Transition fitting comes in various shapes and is used to connect pipes made of different materials. Transition fittings are used to join plastic pipe to galvanized, copper to plastic, and so on.

fixed skylight SKYLIGHT whose glass or plastic light is fixed.

fixed window A WINDOW with no moving parts, also called a stationary window.

How fixed design opened up

Once, the great majority of windows had plain, rectangular designs. As the energy crisis of the Seventies came along, windows drew hard scrutiny as energy wasters. As an outgrowth of the attention to energy conservation, window design received attention. One of the results was the fixed window. With no operating parts, the fixed window comes in a wide variety of designs: trapezoid, half circle, circular, triangular; the list goes on.

Fixed windows were created to increase the amount of light in a home, but they also provide natural design interest. They are often intentionally combined with other operable windows to create interesting designs, some quite beautiful.

fixture 1. The sink, toilet, or tub. **2.** Electrical devices such as lights and water heaters. Such items are called fixtures because they are considered permanently placed in a building.

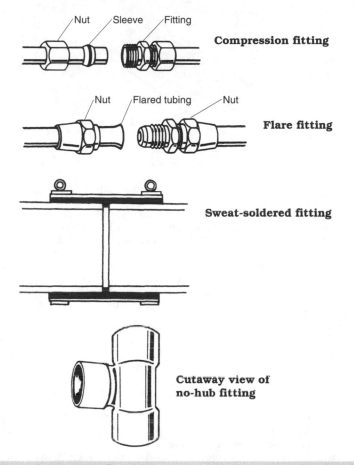

Compression fitting

Nut   Sleeve   Fitting

Flare fitting

Nut   Flared tubing   Nut

Sweat-soldered fitting

Cutaway view of
no-hub fitting

With modesty came fixtures

Bathing and going to the bathroom have not always been viewed as private activities. Until the dawn of the 20th century modesty was not necessarily a virtue. Bathing was a communal or familial activity. In Europe, monarchs granted audiences and women received suitors while seated on the toilet. And in America, the walk to the outhouse announced the person's purpose.

In Victorian times, Americans started to view personal hygiene as private, a need answered by bath fixtures. Unfortunately, it was need for privacy rather than an awareness of the bathroom as an important room that led designers to cram all three fixtures into a 5′ × 7′ space.

flagstone Flat stones, from 1″–4″ thick, used for rustic walks, steps, and floors.
Flagstone used for floors is split into manageable slabs. The most common flagstones used are SLATE and BLUESTONE. *See also* STONE. *See illustration on following page.*

flange 1. Top and bottom longitudinal members of a BOX BEAM **2.** In plumbing, the metal disc that the toilet is mounted on.

flapper ball Rubber valve that covers the flush valve in a toilet tank.

LEEDS COLLEGE OF BUILDING

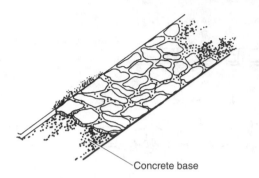

Flagstone

Concrete base

flashing Metal strips used to seal off joints at roof intersections and projections. Many roofers back up flashing with a self-adhering bituminous membrane because the roof is most vulnerable to leaks at these points.

   Aluminum flashing is widely used for step FLASHING, which is usually covered by roofing and siding. Aluminum is not highly regarded for use in other ways. It can't be soldered, something called "tinning the iron," like other flashings, and it often oxidizes and pits in polluted or salty air. Nor is aluminum flashing good for use around masonry; the acids and lime in the masonry can eat the metal. If aluminum must be used, the 0.032 thickness flashing lasts longer than the thinner versions, and painted versions last longer than unpainted ones.

   Copper flashing comes "cold rolled" or soft. Because copper is easily worked, roofers prefer it for decorative jobs and for complicated shapes like the tops of VALLEYS, where malleability is an asset. Copper eventually develops a green patina, which can result in stains on siding and trim. When rain runs off the flashing, the copper is effectively being etched by acid. If this occurs, there will likely be bright red streaks on the roof and green stains on the siding.

   Although galvanized flashing is the least costly, the protective zinc coating, which does not rust, eventually wears away, letting the steel rust. This can be seen in old sheds with galvanized flashing.

   Lead, which is malleable and durable, is a favorite material for CAP flashing. It is also commonly used at the tops of VALLEYS and at the bases of DORMERS and CHIMNEYS. Lead is paintable and will not stain.

   Lead-coated copper has the same working characteristics as cold-rolled copper. Roofers like it because it does not stain and is easy to work.

   See also STEP FLASHING. *See illustration on following page.*

flash point The temperature at which a particular material will ignite. Flash points differ greatly among products. For example, paint thinner has a flash point of over 100 degrees. In other words, it must be at least 100 degrees before a spark will ignite it. But acetone has a zero degree flash point, making it one of the most hazardous materials to use.

> **A most dangerous product**
>
> How dangerous is acetone? Very. One woman was cleaning her fiberglass tub with acetone one day and as she did, the fumes built up. When a phone rang, the bathroom blew apart. The spark from the phone ringing was all that was required for ignition.

flash set Quick hardening of MORTAR.

flat roof A perfectly flat or slightly pitched roof.

   Flat roofs are quite different from pitched ones. Because they are flat, they must be strong enough to take, in certain areas of the country, much snow and ice. And because water does not run off as readily as on a pitched roof, it can collect. If there is the slightest gap in the roofing material, the water can

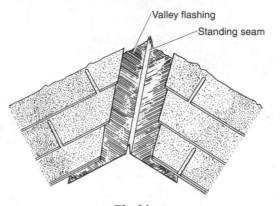

Valley flashing

Standing seam

Flashing

get in. For this reason some slight pitch is usually desirable, or a CANT STRIP is installed. *See also* ROOF.

flat skylight Skylight that is flat rather than concave. Glass is the only type of skylight that is always flat.

flier A simple kind of TREAD. Steps vary in complexity of design; fliers is one of the simple ones.

flitch girder Consists of double JOISTS sandwiching a metal plate at least ¼″ thick, the assembly held together by bolts.

float Process of spreading PLASTER, STUCCO, CONCRETE, etc., on a wall. A tool called a float is used for concrete. *See also* SKIM COAT.

float ball Part of flush tank that helps control water flow.

floatstone A stone used by masons to smooth gauged brickwork. Also known as a rubbing stone.

floorboard Component in a finished floor.

A number of different woods are used to make floorboards, including maple, oak, and hardened pine. The key is that the wood be hard enough to stand up to traffic.

Board edges are machined in a variety of ways for joining such as square-edged, rabetted, TONGUE-AND-GROOVE end-matched, which means tongue-and-grooved on both sides as well as the ends. *See also* STRIP FLOORING *and* PLANK FLOORING.

floor plan *See* CONSTRUCTION DRAWINGS.

flue Space or passage in chimney through which smoke, gas, and/or fumes rise.

A well-designed and well-built flue is a critical element in CHIMNEY/fireplace design. If the wrong design is used, it can lead to smoke and fumes not passing out of the house, or the fire going out.

flue lining Fire clay or terra cotta pipe either round or square used to line a CHIMNEY. *See illustration on following page.*

fluorescent A type of lighting fixture. Known as a "lamp" in the trade.

flush 1. Perfectly even with something. Getting something flush is one of the most common building goals. One trims boards flush, draws off plaster flush, draws down a toilet FLANGE flush with the floor. **2.** A flat, seamless panel of some sort, such as a door.

folding door Door with multiple sections hinged together.

folding stair Collapsible STAIRS that lead to and fold into the ceiling when not in use. There are catches on the sides of stringers that lock the stairs in place. The underside of the stairs is often finished so the stairs are as invisible as possible when mounted in the ceiling.

foliated Surface decorated with a leaf design.

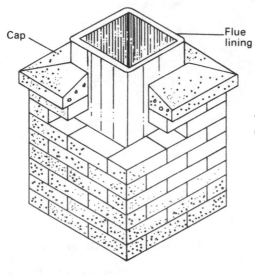

Cap

Flue
lining

**Top of a chimney
showing flue and cap.**

footcandle One way to measure light. Footcandles measure the amount of light that falls on a surface, all parts of which are at a distance of one foot from a light source, a candle.

> ***Footcandles a problem***
>
> Some bathroom designers note that specialists in lighting often make their recommendations for particular baths based on footcandles. The problem, these designers say, is that the footcandles cannot be measured accurately until the job is finished; there are just too many variables. Most designers give their recommendations in terms of WATTS or LUMENS. *See* FOOTLAMBERT *and* CANDLEPOWER.

footing Broadened concrete base supporting a FOUNDATION wall.

The name comes from the fact that footings are the lowest house members, forming a base for the foundation. Weight loads on the foundation transfer to the footings, which in turn superimposes the load to the soil.

There are various types of footings, but poured CONCRETE footings are used the most. Whatever type is used, it is good building practice to locate the footing at least 1½′ beneath the frost line (as deep as 8′ in northern sections of the USA).

Footings are needed because houses settle. How much they settle depends on the type of soil they're built on. Even though houses generally do not settle much (½″ to ¾″), this can still result in substantial cracks in the foundation wall.

Footings work by spreading the load of the house. The narrower the foundation wall, the more pounds per square inch is exerted. Footings, being wider, distribute the load better and make it less likely that the earth below will give a little, just as someone wearing a standard flat shoe exerts less pressure on the earth than someone wearing a spiked heel.

footlambert A way to measure the amount of light reflected. Engineers describe it as the uniform lighting of a perfectly diffusing surface that reflects light at the rate of one LUMEN per square foot. The number of footlamberts is measured by counting the number of FOOTCANDLES reaching the surface and multiplying this figure by the amount of light reflecting off the surface. *See also* CANDLEPOWER.

Not all the soil may settle, but the footing is always made to handle the worst possible scenarios. If most of the house exerted a load of 2000 pounds per square inch, but another part exerted over 3000 pounds, the footing would support 3000 pounds.

forced-warm-air system One type of heating system for a building. *See also* CENTRAL HEATING SYSTEM.

Formica Most common brand name for PLASTIC LAMINATE. Also known as, 'Mica.

forms 1. The wood or other assemblies into which concrete is poured and allowed to harden to get its shape. **2.** Shapes of various building components into which concrete is poured and retained until hard. *See also* EPS FORMS.

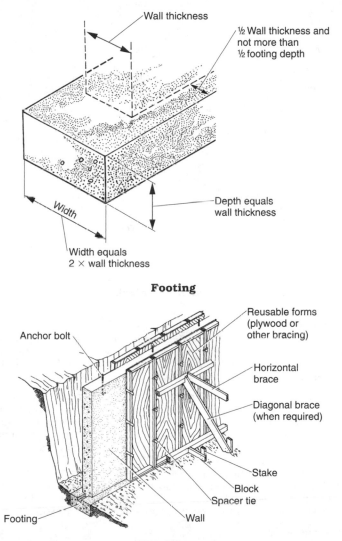

Wall thickness

½ Wall thickness and not more than ½ footing depth

Depth equals wall thickness

Width

Width equals 2 × wall thickness

Footing

Anchor bolt

Reusable forms (plywood or other bracing)

Horizontal brace

Diagonal brace (when required)

Stake

Block

Spacer tie

Footing

Wall

Wood forms

foundation The masonry or wood structure which form the base of a building.

The foundation is the supporting structure that bears the load of the weight of a building. There are three basic kinds of foundation: Crawl space, full, and slab.

A crawl space is a foundation with shorter walls than a full foundation. Like a full foundation, it consists of a footing topped by a foundation. *See* CRAWL SPACE.

A full-four foundation is concrete or masonry walls built high enough so that when the house is completed, a basement is available for use.

A slab is a deck of concrete that rests on earth and upon which all framing members rest. *See also* FOOTING.

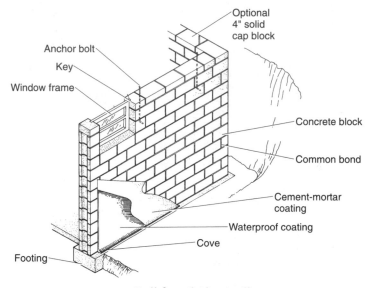

Full foundation wall

> ### *Foundation offering*
> Primitive peoples once offered the lives of young virgins as payment to assuage their gods to ensure that a building would stay stable. A remnant of this practice is the superstitious habit some builders have of burying some coins in the foundations of buildings they erect.

foundation vent Screened or otherwise covered opening in a foundation wall. Vents are usually screens that can be opened or closed as required. Like their counterparts on the roof and elsewhere, vents are needed to insure that warm, moisture-laden air is bled off to keep condensation at a minimum.

fountainhead A solid-surface material with a solid color and a soft, textured stone look. It is used on COUNTERTOPS and in other areas where solid-surface materials can be used.

frame 1. The framing or wood skeleton of a house, also called frame construction. **2.** To erect the framework for a house, window, or door. See also BALLOON FRAMING, PLATFORM FRAMING, POST-AND-GIRT FRAMING.

> ### *In the beginning*
> The first kind of framing utilized for houses in America was simplicity itself. After staking off a small rectangle of land, the builder would dig a hole at each of the corners and perhaps between the corner holes.

A post—either a tree trunk or a beam hewn from one with a broad ax—was mounted in the hole and held plumb with stones and earth. Beams were connected to the tops of the posts, and rafters were notched into these crossbeams to create a framework. Unfortunately, early settlers knew little about the effects of insects, rot, and water in the bottoms of posts, and many posts rotted away.

The next development was in braced-frame construction, also known as "barn eastern" and "combination" framing. This framing was descended from the heavy timber framing that had been used in Europe for hundreds of years. Circular saws and other power saws were, of course, non-existent and so were machine-cut NAILS. As a result, braced-frame construction involved as little sawing and nailing as possible. Heavy timbers characterized this type of construction.

In the 1800s, braced-frame construction shrank. Beams grew narrower and shorter because power saws came to be, and nails became plentiful.

As in BALLOON and PLATFORM FRAMING, anchor bolts secured a sill plate to a foundation. Instead of 2×4s, it was composed of $4' \times 6''$ members joined together at the corners with half-lap joints.

The posts, each $4' \times 6''$ or a massive $6'' \times 8''$, were spiked to the corners of the sill. The posts extended to the top of the house like balloon framing, held in place by temporary bracing.

Girts, horizontal members installed between post at each floor level, were next. These were nailed in place and secured with MORTISE AND TENON joints. Because of labor costs, much braced-frame construction today does not use mortise and tenon joints; instead, it is supported with blocks of woods spiked to posts.

There are two types of girts. In "dropped" girts, two girts are placed at opposite sides of the building to support the joists. Raised girts are also set at opposite sides of the building, but they support loads from above.

In early versions of braced-frame construction, girts could support the joists and all the loads coming from above. As time went by and the members narrowed, it became necessary to install studs beneath them for supports.

Studs are installed next and these may be 2×4s or 2×6s. On top of them go the plates, either individual $4'' \times 6''$ members or two separate members of the same combined thickness. The method involving two members is less likely to result in warping. Whichever, the plates are joined with the post with half lap joints like the sill.

framing fastener Metal shape with pre-punched nail holes used with nails or screws.

A number of companies make these fasteners. Made of either 16- or 18-gauge zinc-coated metal, the device is set between a pair of framing members. Nails or screws are driven through the holes to lock the members together. Using fasteners makes framing faster than using nails alone.

Framing fasteners also strengthen connections. A number are designed to join members to resist hurricane-force winds. It should be calculated first whether extra strength is needed; certainly, framing fasteners cost much more than using just nails or screws. Additionally, not all BUILDING CODES accept these fasteners.

The joist-and-beam hanger, a squarish U-shaped FRAMING FASTENER, is used to hang beams or joists of various sizes. These hangers save time. In standard JOIST hanging, for example, the joists must be notched, and then rested on and nailed to a LEDGER strip, which is nailed to the box of the perimeter joist. The framing fastener is simply set in and nailed in place on the box joist.

The post-and-beam connector is a T-shaped connector for installing horizontal beams on posts. To use it, nail the fastener to the post, flip the horizontal members into it, and nail in place. No toenailing is needed.

A post holder is a U-shaped framing fastener for setting posts in place. The post slips over a bolt projecting from a concrete foundation; once on the post, the bolt locks in. It has the added advantage of keeping the end of the post out of direct contact with the concrete.

The right-angle fastener is a rectangular piece of metal bent at a right angle. It can be used to ensure that studs are aligned.

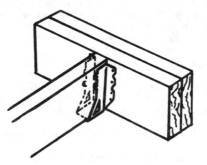

One type of framing fastener. This one accepts bottom of post.

This framing fastener is designed for hanging joists.

Barbed plate-type framing fastener is commonly used on trusses.

A truss fastener is a flat metal plate with holes for nailing through or barbs to hammer into the truss. This fastener eliminates the need for GUSSETS; it is also useful in connecting horizontal and vertical framing members. To use the nailing type, position the plate so that it overlaps the members to be joined and then simply nail it on with the nails provided. The barbed type is similarly secured, but sometimes a special tool is needed.

freestone Non-specific stone that can be easily worked with masonry tools without splitting.

French door Door with small multiple windows.

French doors are available for between rooms as well as to serve as PATIO DOORS. They swing out, being hinged along one side of each door. French doors may consist of two movable doors, or four—two fixed and the movable pair hinged to them.

It was once felt that French doors were not as secure as sliding doors for patios, but this has changed since three-point locking came into effect. Now the doors can be locked at the top, at the threshold, and at the midpoint to each other.

French doors are normally available only in wood.

frieze Horizontal member connecting the top of the siding with the soffit of the CORNICE. Through the years, the width of the frieze has varied from narrow to wide. American Colonial style houses contained friezes wide enough to install windows in.

frog Depression or groove in the surface of a BRICK. The frog will be in one or both of the larger sides of the unit, and provide a key for mortar joints while saving money at the same time.

frost heave The swelling and bulking up of the ground because of water that turns to ice and expands. One often sees heaving of sidewalks, driveways, and the like where the material installed on the ground does not go deep enough to get beyond the FROST LINE. When the water in the ground turns to ice, it enlarges and pushes up against the concrete or asphalt, resulting in cracking and failure of the material. The condition can also be seen when tree roots bulk up; they are heaving.

frost line In winter, the level above which the ground is frozen. This line varies greatly depending the area of the country. In the south, it might be only a few inches; in the colder western states, the frost line might go down a few feet. Regardless, the frost line is an important building consideration, as evidenced when something heaves. *See* FROST HEAVE.

FRP Fiberglass reinforced plastic, a tough, non-scuff coating for plywood used on plywood intended for forms and other uses.

full mortar bedding Method of applying MORTAR over the entire surface of a concrete BLOCK. This type of mortar is normally used on FOUNDATIONS and other load-bearing situations. *See also* BED, FACE SHELL, *and* MORTAR JOINT.

fungi Microscopic plants that live in damp wood and cause mold, stain, and decay. *See* MILDEW.

furred A wall or other surface made level by securing thin wood strips to it. *See* FURRING.

furring 1 × 2 boards used to level walls and ceilings prior to application of some other materials.

Walls are rarely perfectly true, which is what gives furring its place. Furring can be used on any surface, including concrete when secured with the proper fasteners. The material is often used with shims, thin pieces of tapered wood. Drive behind the strips with a hammer when making them true or if a level is required. *See illustration on the following page.*

fuse Safety device in an electrical system.

Although CIRCUIT BREAKERS have largely replaced them, fuses are still used in homes today. Like circuit breakers, fuses are designed to handle a

Furring

Furring

specific amount of CURRENT. When the current exceeds that amount, a metal linkage in the fuse melts. The current stops flowing before the overheated wires can heat up. A fuse is a safety device and should never be circumvented. A plug-in is the classic type of fuse. It has a threaded end and a small window through which one can see if the linkage inside has melted (the fuse has "blown"). Plug-in fuses can handle from five to 30 AMPS. Fifteen amps is normal for lighting circuits, while larger appliances use 25 and 30 amps.

A type S plug-in fuse contains two parts: one that looks like a standard plug-in fuse, and the other like a threaded adapter. The advantage of this type of fuse is that it is tamper-proof. Standard plug-in fuses are interchangeable no matter the amperage capacity; they can use a higher amperage than is safe. Type S fuses cannot be compromised. The adapter accepts only one level of amperage—a 20 amp fuse cannot, for example, be screwed into an adapter for a 15 amp one.

Cartridge fuses, a shotgun-shell shape with metal caps on the ends, are used less in the home than plug-in types. They are designed for handling power demands above 30 AMPS and 220 VOLTS for such things as appliances and air-conditioning equipment, where a separate fuse is required to protect a particular appliance circuit. There are two range sizes: 10 amps to 30 amps and 35 amps to 60 amps.

Fuse

gable The portion of the ROOF above the line of a double-sloped roof. Gables occur at the ends of a building. The standard gable end is two straight peaked sides, but there are other style buildings that are different to some degree although they have the same basic gable shape.

Gable: gable-style house

gable roof Roof of a GABLE-style house.

gable vent Screened vents or louvers in the gable ends of a house. Vents are used to bleed off warm air that otherwise might collect and condense, doing damage.

galvanic action When two dissimilar metals come into contact in the presence of moisture, the more active metal corrodes by transferring ions to the more passive; the more passive metal remains unharmed.

Galvanic action takes place to varying degrees. With the following metals—zinc, aluminum, steel, cast iron, lead, tin, and copper—the further away the first metal is from the last the better. Putting zinc and copper together would be a worst-case scenario.

113

The best way to prevent galvanic action—also called *galvanic corrosion*—is to keep the metals apart. Failing this, they should be separated with some sort of BUILDING PAPER.

galvanized steel Steel that is coated with zinc to make it weather-resistant or weatherproof. Only items that are called "hot-dipped galvanized" are relatively weatherproof.

garret A small room in the same area as an attic. A garret is associated with writers and other artists who sought the solitude of such places to help their creative muses. "Garret" comes from the word "guard," and many years ago signified a watchtower.

gasket Any of several soft fibrous materials used on the internal working parts of faucets, valves, and the like.

Gaskets are used because mating hard metal parts will not stay in place, and a gasket gives rather than resists, placing less stress on the metal.

gel coat Final finish on a variety of plastic FIXTURES. There are various types of gel coats, but all share an ability to be repaired simply if scratched or marred.

GFCI Ground fault circuit interrupter. A GFCI is a safety device that protects the user against electrical shock. If there is a slight stray of current leakage (just $^5/1000$ of an AMP), the GFCI will sense it and cut off the flow of electricity through the circuit.

GFCIs come in three different forms. One looks like a standard receptacle and is designed to be used as a receptacle. Another is a combination GFCI/circuit breaker and is installed in the circuit breaker box. The third type is portable. They are available to protect 15 and 20 amp circuits.

GFCIs are particularly important—and in many localities required—where water is used, such as on a line that goes to a pool, or in a bath or in a kitchen. There is normally a small amount of electrical leakage in house wiring, which can mistakenly trigger a GFCI/circuit breaker; hence, many people opt to have a receptacle GFCI installed where it is needed.

See also GROUNDING.

Why people are electrocuted

In America, around 150 people a year get electrocuted in the home. Normally, these electrocutions do not occur while they are touching something with only one hand. Unless it is wet, a hand is a poor conductor of electricity. The electricity would have to flow to the ground through the person to complete the circuit—an unlikely event.

Far more dangerous is when a person touches an electrically hot side of something with one hand, and has the other hand on metal. The current goes through one hand, across the arm, across the heart, and then across the other arm. The heart operates on electricity, beating when it receives an electrical charge that is about 0.06 amps; but the current that hits it is too great for it to bear, and the heart stops—cardiac arrest.

GFCI

gingerbread A general description for fancy trim on a house.

girder A large or principal beam of wood or steel used to support loads at points along its length. The ends of a girder usually rest on the foundation and/or there are PIERS along its length.

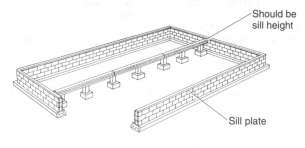

Should be sill height

Sill plate

Girder

girt In old-time POST-AND-GIRT FRAMING, the name of a horizontal supporting piece. Depending on location in the structure, girts were given various names, such as front girt and rear girt.

glass Transparent hard sheet material.

A variety of glass types are available, depending on one's needs. Following is a roundup:

- **Single strength.** This is $3/32''$ thick and comes in a variety of standard sizes or can be cut to order.
- **Double strength.** This is standard single-strength glass but $3/16''$ thick—twice as thick as single strength. This glass is good for cellar windows or wherever there is a possibility of the glass' being hit by an object.
- **Single strength glass.** Used for ordinary windows in a home.
- **Plate.** This comes $1/4''$ thick in various sizes and with beveled edges.
- **Plate glass.** Commonly used on coffee tables, countertops, and other surfaces. It can be cut the same way as other glass can be cut.
- **Tempered.** Looks like regular glass. When it breaks, it breaks into small, relatively dull rather sharp-edged pieces. All tempered glass is custom ordered; it is "drawn" to the proper size. Once made it cannot be cut.

glass block Building block made of solid glass.

In the last decade or so, glass block, once relegated to the basement where as windows they allowed light in and provided a measure of security, have come on strong. Today they may be used anywhere in the home and in other building situations for their aesthetic value.

A variety of standard sizes are available: $4'' \times 8''$ (3.6 pounds), $8'' \times 8''$ (6.4 pounds) and $12'' \times 12''$ (15.3 pounds). All of these blocks are $3\frac{7}{8}''$ thick, with some half sizes available. There are also "thinline" blocks, which as the name suggests are thinner: $3\frac{1}{8}''$ and anywhere from 3 to 5 pounds lighter than the standard type. Glass block are not load bearing.

Blocks come in a variety of colors and finishes from crystal clear to ones having snowflake images impressed in them, and in smooth and textured finishes. A variety of shapes are available, including curved block for curved wall sections.

The classic way for building with block is with a special kind of mortar used for the block. Doing the job is not simple; it is much slower than erecting a brick or block wall, mainly because the MORTAR JOINTS must be perfect.

Manufacturers also make systems mainly geared to the do-it-yourself market, where block are assembled with frames and silicone.

glaze coat **1.** Top layer of asphalt in a built-up roof assembly. **2.** Interim application of a thin coat of asphalt to a roof when final finishing is delayed. **3.** Final coat in paint finishing process.

glazier's points Small metal fasteners used to hold glass in its framework. These come as flat metal triangles or a V-shape with a raised portion. Glazier's points are pushed into the frame and against the glass, locking it in securely. *See also* GLAZING COMPOUND.

glazing compound Material used to seal the joint between window glass and the frame. This is commonly called PUTTY, but glazing compound is different, chiefly in that it will stay flexible, and therefore is not as likely to dry up as putty. *See also* GLAZIER'S POINTS.

gliding window Window with short vertical panels which slide back and forth, like a PATIO DOOR.

glueline **1.** The point at which adhesive is applied. If two boards were glued together, the glueline would be at the edges where they join. **2.** The adhesive joint formed between veneers in plywood panel.

glulam Beams formed by gluing layers of strong woods together. Glulam is short for glued-laminated structural timber. End jointing and edge jointing permit production of longer and wider structural wood members than are normally available.

Glulam lumber members are thicker than other types of laminated lumber (*see* LVL) and differ in look. It is made from solid lumber, and as such can be finished and left in the open while others may not. Glulam is also called an "engineered wood."

Glulam

grab bar Bar installed in the bath tub so it can be gripped for safety by someone bathing.

grade **1.** The ground level around a building. The natural grade is the original level. Finished grade is the level after the building is complete and final grading is done. "Above grade" is above this level, and below it is "below grade." **2.** Various qualities of LUMBER and PLYWOOD and other materials.

grain The natural growth pattern in wood. Wood grows in many ways with various characteristics in the grain, all of which affect performance and appearance. Following is a lineup:

  • **Close-grained.** Wood with narrow, inconspicuous annual rings. Often said to describe wood with small and closely spaced pores; also known as "fine-textured" wood.
  • **Coarse-grained.** Wood with conspicuous annual rings. The term is sometimes used to describe wood with large pores, such as oak, walnut, ash, and chestnut. Also known as "coarse-textured" wood.

- **Cross-grained.** Lines of grain depart from running roughly parallel to the edges of a board and are either diagonal or spiral or a combination of the two.
- **Curly-grained.** Wood with a grain pattern distorted to the point that the wood has a "birdseye" appearance. The areas showing curly grain may vary up to an inch in diameter.
- **Diagonal-grained.** Grain runs lengthwise in trees and is strongest in that direction. Similarly, grain usually runs the long dimension in the face and back veneers of a PLYWOOD panel, making it stronger in that direction. It is suggested that structural wood panels should usually be applied with the long dimension perpendicular to or across supports.
- **Open-grained.** Wood with particularly large pores, such as oak, ash, chestnut, and walnut.

grain raise When wood fibers swell and rise above surrounding wood. This is often caused by water, which swells the wood, raising the fibers. The solution is to use a fine sandpaper and sand the fibers off.

granite A very hard natural stone.

    Granite is extensively used for countertops, flooring, and other tasks in the house. It consists of a variety of hard minerals. Granite has an extremely low absorption rate and is not subject to staining, and is less susceptible to scratching than other materials. It also has a coarse grain that makes it more slip resistant than MARBLE.

gravel Small stone of various sizes. Gravel is a key component in building. It is an ingredient in AGGREGATE, forms a wearing surface on FLAT ROOFS, and is important in many jobs that involve draining away water. While it is strong enough to support heavy weight, it also drains well.

gravel stop A flanged device used to keep gravel from washing off the edge of a roof.

greenboard Water-resistant DRYWALL. This type of drywall gets its name from the green paper covering it. Greenboard is not waterproof, but it is more water-resistant than standard drywall, and is therefore commonly used in the bathroom under tile. It is not used on ceilings—it is too brittle. *See also* BLUEBOARD *and* CEMENT BOARD.

green brick Brick that has been formed but not burned to make it hard enough for use.

Greenfield *See* CONDUIT.

green masonry Uncured CONCRETE MASONRY units.

green mortar Occasionally used to describe MORTAR in a state prior to setting.

green wire In electrical installations, the wire that is used as the GROUND.

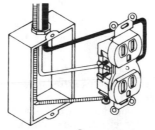

Green grounding wire

Neutral (white) −

Line +

Green hex screw

Green wire

green wood Wood that contains a high water content prior to air or kiln drying.
grounding The method of connecting all electrical facilities to the earth.

It is crucial to understand grounding to ensure a safe, well-designed electrical system. To understand grounding, it is best to forget the normal electrical CIRCUIT. The ground is simply a separate metallic path for the electricity in case a device such as a toaster or other appliance develops an electrical leak. If a leak occurs and there is no grounding system, electricity could electrify an appliance. The user would touch it and get shocked. Fortunately, metal is a much better CONDUCTOR, or carrier, of electricity than a human body, so electricity will take that path rather than the person.

There are various ways of grounding a system out. One way is to attach grounding conductor to water pipes to provide a route for the electricity; another is to run a ground wire to a copper rod buried in the earth.

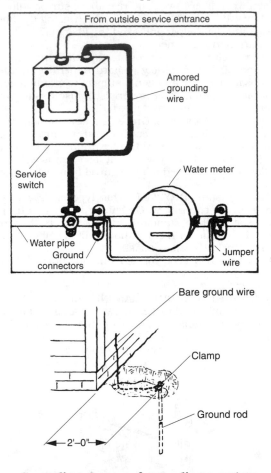

Grounding: 2 ways of grounding a system

grounds 1. 1″ × 2″ strips of wood nailed around windows and doors to serve as a guide to make PLASTER level with the molding or BROWN COAT in a plaster job. The strips on the bottoms of the studs are also used as a nailing surface for molding. Grounds are important in a plaster job; without them, the plasterer has no precise guideline as to just how thick the brown coat should be. **2.** Nailing strips on masonry walls for attaching trim or furring to.

grout **1.** A cementitious material used to fill the gaps between CERAMIC TILE. Grout comes in a variety of types, as follows:

- **Non-sanded.** This grout does not contain sand. It is used for tiles that require small joints (⅛″), and on soft glazed tile that could be scratched. Non-sanded grout looks smooth and is available in a wide variety of colors.
- **Sanded.** Sanded grout has a rougher appearance than non-sanded, and is used where joints are wider (up ½″). One problem with the material, because of its absorptive qualities, is staining. Using a latex additive known as MILK instead of water can alleviate this problem greatly.
- **Epoxy.** Made by mixing grout with a hardener. The epoxy bonds tenaciously, has great impact resistance, and is highly resistant to chemicals, though the grout can be stained. Epoxy grout requires careful installation.
- **Silicone rubber.** In situations where a flexible grout is desired, silicone works well. Once cured, the grout resists stains quite well.

Installation of grout varies from type to type, but it essentially involves spreading the material over the tile, forcing it into place with a squeegee, and wiping the tiles down. Curing times can vary.

Grout comes in a wide variety of colors, and with or without MILDEW-CIDES. It is better to get grout with the mildewcide because the bath is an ideal breeding ground for MILDEW.

2. A mixture of PORTLAND CEMENT, sand that is poured into the CORES when CONCRETE BLOCK is being reinforced. On walls where RE-BAR is used, the grout mix, which is made thin enough so the aggregate will not separate from it, is mixed and poured into the cores where the bars are located. They are PUDDLED, allowed to set a bit, and puddled again. Normally, walls are built in 4′-high sections; bars are 4′ long. Hence, adding grout as described above (it is called "low-lift grouting") is required at 4′ intervals.

Grout: grouting with a bucket

120   grouted masonry • gutter

grouted masonry MASONRY in which the air cells of BLOCK are filled with grout as the work progresses.

gusset A flat piece of lumber, plywood, or the like used to connect wood members. Most commonly used in joints of wood TRUSSES. Gussets may be applied to one or both sides of the joint. Also called a gusset plate.

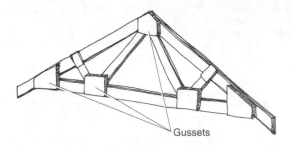
Gussets

Gussets

gutter Trough-like product for carrying water away from the house. In profile, gutters may have a half round or K-shape and be 4″–5″ across at the top. The 5″ size is necessary for most areas.

Gutter comes in 10′ lengths, sometimes 20′. It is made of a variety of materials, each requiring "fittings"—downspout tube, leaders, and hangers. Fittings vary slightly from one manufacturer to the other, but outlets that carry Brand A gutter will carry the companion fittings. Hangers may be the concealed type (they cannot be seen by looking at the gutter); a type that wraps around the gutter and fastens to the roof rather than to the fascia, the way the concealed type does; or a type designed to fit around the molding of the house.

Aluminum spikes and "ferrules" may be driven through gutter into the fascia and, preferably, into the ends of the rafters. Lifting up the ends of the shingles on the house ordinarily shows where rafter ends are.

Following is a look at the various types:

- **Aluminum.** Available in various configurations and in 10′ lengths, and up to 32′ long from factories that specialize in making gutter. 10′ lengths are 0.027″ thick, while longer lengths are 0.032″. (Some companies sell gutter thinner than 0.027″, known as "Reynolds Wrap" in the trade. It is available in a wide variety of colors. "Seamless gutter" is aluminum gutter extruded from a machine to exact lengths needed and without joints. Seamless gutter is the best one can buy. Because it has no joints, it is not subject to the same leaks as jointed gutter, and it is commonly a heavier gauge metal. Seamless gutter must remain rigid in the long lengths it is extruded in. It is available in many more colors than standard aluminum gutter.
- **Copper.** Copper gutter comes in standard shape but is made of copper. It is available in 10′ lengths and is very expensive. After copper gutter weathers, "verdigris" (a greenish colored compound) may run down and stain other surfaces. Installing copper requires that joints be soldered.
- **Galvanized steel.** Available in standard gutter shapes and in various baked enamel finishes. Galvanized steel gutter is available in 10′ and 20′ lengths.
- **Vinyl.** This is available in standard gutter shapes but in a limited range of colors. It is made of polyvinyl chloride and comes in various lengths: 10′, 16′, 21′, and 32′. Like all plastics, vinyl gutter is not

subject to rotting, or peeling, but it can crack from expansion and contraction in cold weather. It is installed like any other gutter but is connected with sleeve connectors in which the ends of sections fit into sleeves.

- **Wood.** Wood has the standard gutter shape, but since it is made of wood (usually fir), it is thicker than the other types. Wood gutter can be used on most homes, but it is expensive and heavy—five or six times the weight of metal gutter. It comes in 50′ lengths. Every couple years, a protective coating is applied to keep it from rotting, and it must, of course, also be painted.

Gutter: number one cause of leaks

Gutters that are properly installed route thousands of gallons of water away from houses. But if they are not installed correctly, or if they leak or clog up from ice or debris, water can spill down, saturate the foundation, and seep into the house. It is safe to say that incorrectly or leaky gutters are the number one cause of damp or wet basements.

Pitching gutter properly

A common mistake made when installing gutter is to pitch the gutter as if the house were level. One snaps a line along the FASCIA with a certain dip every 10′ or so and installs the gutter with the back edge aligned with the line.

The problem is that houses are rarely level. To ensure proper pitch, a level line must be struck independently of the house's position and another line struck for pitch off that first line.

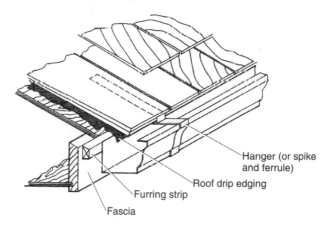

Hanger (or spike and ferrule)
Roof drip edging
Furring strip
Fascia

Gutter: gutter hanger

gutter spikes Six-inch long NAILS used to secure metal gutter to a house. Gutter spikes come in two versions: aluminum and plain steel. Aluminum gutter spikes, being light, are used with ferrules. The nail is slipped through a ferrule before being hammered in place. With a steel nail there is less danger of splitting the wood.

gyp board Lingo for DRYWALL.

gypsum Natural rock used in the making of DRYWALL and PLASTER, among other things.

Gypsum is a mineral in rock form and, as mentioned, it is the basis of plaster and drywall. Although it also occurs in powder form, gypsum is a rock that is mined and put through a process that involves boiling it to remove its

high water content (usually over 20 percent), purify it somewhat, and reduce it to a powder. In this form, water can be added to it and gypsum becomes plastic and workable, a great building material. It is plastic and
workable. The original gypsum, mined from under Paris in the late 1900s—
and called plaster of Paris, was not workable for long, but today chemicals
can be added to gypsum to slow its return to its rock-like state considerably. When this occurs, gypsum has a high water content again, and is essentially a fireproof material because of its water content.

The Egyptians were using gypsum 5000 years ago, and based on the
huge stores of it in the United States there is no reason to believe it won't
be around for another 5000 years.

gypsum lath Gypsum board with holes in it through which plaster seeps and
forms KEYS. *See also* PLASTER.

H-clip Metal clip into which adjacent edges of plywood are inserted to hold them in alignment.

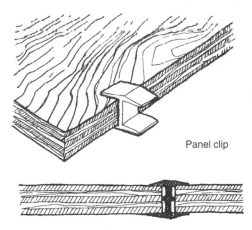

Panel clip

H-clip

half bat Half a brick.
half by Used to describe any lumber that is ½″ thick.
handed *See* HINGE.
hanger bolt Metal shaft threaded on one end like a LAG SCREW and having machine threads on the other.

123

Hanger bolts are ordinarily used for connecting parts of wooden furniture but are also used for hanging things. The lag screw portion may be turned into a beam and the machine-screw part threaded into the matching threads on the fixture or used with a nut and washer to retain anything with a mounting hole.

hangers Upper cabinets in a kitchen.

Hanging KITCHEN CABINETS must be installed on studs—they are too heavy to fasten only to walls—so the stud locations must be marked out on the ceiling or some other place before installation begins so they can be referred to as the job progresses. In most cases, STUDS will be 16″ on center, but in older homes they may be random distances. A variety of tools are available for finding the studs.

If there is a high BACKSPLASH, the cabinet can be hoisted up and have its bottom edge rest on it. The cabinet is placed in approximate position (it is a two-person job, usually) and a ten-penny nail is driven through the rail in back into a stud to hold it there. SHIMS are inserted to make it PLUMB in two directions: the face and the side. When it is plumb, screws are driven through the back of the cabinet into the studs. The adjacent hanger is lifted into position, and if the front STILE edges do not meet perfectly, the carpenter will scribe and trim them after getting it plumb.

This procedure is followed around the kitchen. If there is no backsplash for helping to hold the cabinets, then a temporary nailer is attached to the wall to rest the cabinets on.

hardboard High density FIBERBOARD.

Available in a number of sizes, hardboard is used for such things as UNDERLAYMENT, drawer bottoms, shelving, and a variety of other things. It comes tempered and untempered; tempered is the harder of the two.

The most famous brand name for hardboard is Masonite, named after its inventor.

It came to him all at once

Masonite was invented by scientist W. H. Mason. Mason was in his kitchen one day pressure-cooking wood chips when an explosion occurred, knocking everything askew and firing fluff—formerly chips—all over the room. But all Mason saw was the fluff, and he had an idea. He gathered it up and mixed it with glue and eventually developed a way to make it into usable sheets.

hardpan A hard, dense layer of clay in soil which is difficult to excavate.

hardware General term for describing hardware needed for a particular job. The hardware for a door installation would consist, usually, of the hinges, doorlock, and needed fasteners.

hard to the line Masonry unit laid too close to the mason's guideline.

hard water Water with high concentrations of carbonate minerals. These minerals form deposits that can clog pipes, and are a factor in considering what type of pipe to use. *See also* BUILDING CODE.

hardwood Any wood from deciduous, or leaf-bearing trees. A common misconception is that all hardwood is hard, but some are softer than softwood; nonetheless, if a wood is classified as hardwood, such as oak or maple, it is considered hard.

Hardwood is generally a much better-looking wood than softwood, and is used for building where the wood GRAIN and color will not be obscured by an opaque coating such as paint. It is much more likely to have a STAIN applied, or a clear coating.

As with other LUMBER, hardwood is graded, particularly when it comes to wood flooring (*see* STRIP FLOORING), which is usually hardwood such as oak and maple. Different groups such as the National Oak Flooring Manufacturers Association grade the wood in different ways, but selection comes down to appearance.

harsh mortar MORTAR that is difficult to spread.

hatbanding Lines of paint on the wall that appear darker than adjacent areas, usually caused by excessive rolling or brushing of certain area or a roller with a nap that's too thick for the surface.

HDO High density overlay. An exterior-type PLYWOOD finished with a resin-impregnated fiber overlay to provide a hard finish that does not need further finishing. The overlay material is bonded to both sides of the plywood as an integral part of the panel faces. HDO is used for concrete forms, cabinets, countertops, and any situation where the material takes punishment.

head Top horizontal piece of door framing. It leads into the jambs or rests on them.

header **1.** A BEAM placed perpendicular to joists, to which joists are nailed in framing for chimneys, stairways, or other openings. **2.** The top horizontal piece in a rough opening for a door or window. Though singular, a header often consists of a sandwich of boards. **3.** Horizontal framing member against which the ends of joists are nailed. **4.** In masonry, brick that overlaps two or more adjacent WYTHES of masonry to tie them together. There are three types. **a.** A "blind" header is a hidden brick header in the interior of a wall. **b.** A "clipped" header, also known as a false header, is a bat situated so it looked like a header. **c.** a "Flare header" is a header darker than bricks in the field.

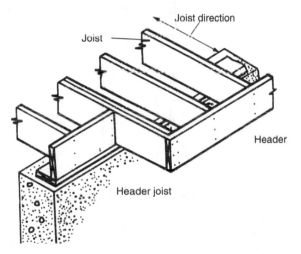

Header

header block *See* BLOCK.

header joist *See* JOIST.

heading course A continuous bonding course of header brick.

head joint Joint between the ends of two masonry units.

hearth The inner or outer floor of a fireplace, usually made of brick, tile, or stone.

heartwood The wood of a tree extending from the pith to the sapwood, the cells of which no longer participate in the life process of the tree. Heartwood may contain phenoloic compounds, gums, resins, and other materials that usually make it darker and more DECAY-resistant than SAPWOOD.

heels Triangular pieces of wood that can be driven into gaps between rough framing and finished items, such as window frames, to provide a solid backing for these items.

herringbone Describes the way a particular material is placed.

herringbone bond *See* BOND.

high lift grouting Technique of grouting MASONRY in one continuous process to the height of a full story.

hinge General term for devices that permit doors to be opened and closed. There is a seemingly endless variety of hinges, relatively few of which come into play for the average home.

Many companies specify that hinges are designed to go either on the left-hand or right-hand side of the door. In some cases, the left is from the perspective of viewing the door opening from the outside of the house; in others, it is from the inside. The truth is, unless you are involved in a special situation, you do not need to know the handedness, because hinges are interchangeable. Turn them over, then put the pin in and they become left-handed or right-handed as required.

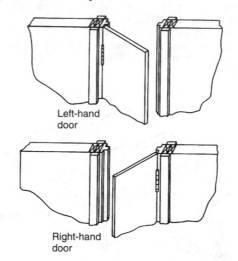

Left-hand door

Right-hand door

Hinge: Handedness in hinges

Hinges come in different sizes. In selecting a hinge, you should know door thickness, weight, and clearance, but a long-time engineering trick makes it simpler: if the hinge seems to be in proportion to the door, use it. For example, if you are hanging an exterior door, you wouldn't use tiny hinges, nor would you use extra-large ones. Generally, if the hinge size looks right, use it. Hinges are many times stronger than they have to be.

Following is a lineup of major hinge types:

• **Butt.** Hinge with two rectangular leaves with screw holes joined by a pin or rod. As mentioned, if a hinge looks in proportion to a door, it likely is. But as a rule of thumb, for a normal weight interior sash door 1⅜″ thick or a hollow core flush DOOR, use 3½″ × 3½″ butt hinges; for a 4″ × 4″ door use one and a half pairs of 4″ × 4″ hinges; and for louvered 1⅛″-thick closet doors use one pair of 3″ × 3″ hinges. Hinges come with "loose" or "fast" pins. Loose pins can be removed and the door taken down. Fast pins cannot be removed and the door taken down, a security plus. Butt hinges are mortised into the door and frame: one hinge and leaf go into a recess in the frame, the others into the side of the door.

• **Butterfly.** Cabinet door hinge with leaves that look like the wings of a butterfly. Butterfly hinges, which are usually brass, BRASS-PLATED, or black wrought iron, are used on FLUSH doors, although there is a variation that can be used on lipped doors.

Hinge: butterfly hinge

• **Double acting.** Hinge with two leaves and knuckles that enable it to turn both ways and fold flat for storage. These hinges are good on shutters and sectional screens.
• **Euro.** This is a heavy, hook-like hinge that is a favorite of contractors because, with a special bit and jig, it is easy to install.
• **H or L hinge.** Hinge in the shape of these letters. It has a black or hammered copper-plated finish and is usually used on early American and Colonial style doors. An easy hinge to install.
• **Pivot.** Cabinet door hinge with bent over leaves pinned together at the top. This hinge, used on kitchen cabinet doors, commonly comes either CHROME- or CADMIUM-PLATED. It is designed to be used with overlay and FLUSH doors only, and is completely hidden when installed.

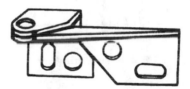

Hinge: pivot hinge

• **Self-closing.** This can be used on all kinds of cabinet doors. It has two straight rectangular leaves with a nylon insert. It comes in various platings. As the name suggests, this hinge is self-closing and as such tends to make the door shut a little too hard. To reduce noise and stress on the door, "dampers"—circles of felt or the like—are secured where the door contacts the frame. *See illustration on following page.*
• **Soss.** Hinge for a FLUSH cabinet door, designed to be completely hidden when the door is shut. It is usually BRASS-PLATED. Soss hinges are for use on doors that are at least 4″ thick. This is a strong hinge, available in five or six sizes, some big enough to use on a standard door. This is a very difficult hinge to install correctly. There is a template available for making matching holes in the frame and door; it is strongly recommended. *See illustration on following page.*

hip The external angle formed by the meeting of two sloping sides of a ROOF.

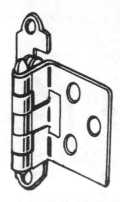

Hinge: Self-closing hinge

Hinge: soss hinge

hip rafters Rafters in a HIP ROOF that extend diagonally from the corner of the plate to the RIDGE.

hip roof A ROOF that rises by inclined planes from all four sides of a building.

Hip roof

holiday A missed spot in painting sometimes known as a "Sunday" or "vacation." A holiday may also be a missed spot when tarring a roof or foundation.

Originally, a holiday was a spot missed when tarring boats. Back in 1785, it was defined in Grose's *Dictionary of Vulgar Terms* as "part of any ship's bottom, left uncovered in paying it."

Today it is quite commonly used by painters to describe missed spots. The term likely arose from the idea of a holiday being a gap of a sort.

Beware artificial light

Painters say that holidays proliferate when one attempts to paint in artificial light. Such illumination just does not allow one to see well, and many a painter has seen this evidence the morning after a paint job at night.

honey Slang for solid human waste. Trucks that haul waste are known as "honey trucks." *See also* SEPTIC TANK.

honeycomb CONCRETE filled with voids. This sometimes occurs when concrete is poorly mixed or not "puddled."

hopper window Window that swings up and down to open and close.

horn Opening in toilet where wastes are discharged.

horsefeathers Fill-in pieces used when preparing a roof for new roofing material. The existing roof TABS will be curled up. The roofer will clip these off, and fill them in with nailing them on. This, then, will supply a flat surface for the new roofing material.

hot stuff Hot BITUMEN.

hot wire In an electrical installation, the wire that carries the current coming into the product.

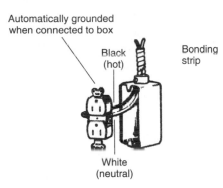

Hot wire

housed string Stair stringers out of which grooves are cut on the inside and into which the ends of TREADS and RISERS are secured. Wedges and glue are often used to help hold the members in the grooves.

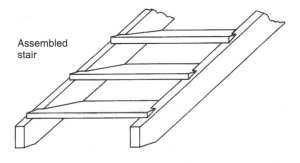

Housed string

humidifier Device for adding moisture to the air. Humidifiers can be valuable to the homeowner, not only in terms of comfort (*see* WINDOW), but also to prevent damage to materials. Just as excessive moisture can damage materials, so can a deficiency. Woods and glues dry out, loosening joints. A humidifier can help end this. *See also* CONDENSATION.

hydrated lime Quicklime treated with water and processed for use. Hydrated lime is the usual material added to MORTAR at a job site.

hydration Chemical process that joins water with cement and AGGREGATE to form CONCRETE, or water and cement to form MORTAR.

hydraulic cement Cement that hardens even in the face of running water. In fact, it is designed to repair cracks and holes in cement with active water leaks. All that needs to be done is to mix the powder with a water into a cementitious slurry, and pack it in and hold it onto the crack with a trowel for a few minutes until it hardens. *See* HYDRAULIC LIME.

hydraulic lime (hydrated lime) Lime that hardens in water. *See* HYDRAULIC CEMENT.

I beam A steel BEAM with a cross section resembling the letter I. I beams are used to support long spans such as basement beams or over wide openings, such as a double garage door, when wall and roof loads are imposed. *See also* LVL.

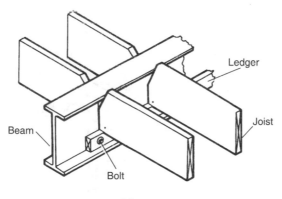

I beam

incandescent A type of light.
incline SLOPE of a roof. *See also* PITCH.
individual appliance circuit *See* CIRCUIT.
initial set First hardening of MORTAR.
inlay *See* VENEER.
insulation Materials high in heat resistance used to reduce heat flow.

131

Insulation works by blocking or slowing down heat flow. In the winter, it helps keep warm air inside the house; in the summer, it keeps warm air out. Insulation saves energy in winter and summer because heating plants do not have to work as hard. The same is true with air conditioning; it does not have to keep displacing warm air coming in from the outside with cool.

Insulation comes in various forms. The key consideration is R-value, how much heat can the particular insulation resist. The higher the number, the better the insulation is at this. Insulation R-values will be stamped on the product.

See also BATTS, BLANKETS, BLOWN-IN INSULATION, BOARD INSULATION, *and* LOOSE FILL.

Ice dams

Snow piles up on the house. As the temperature warms up, the snow melts and runs down the roof, but when it reaches the colder lower portions of the roof, such as above the eaves, it starts to turn to ice. The ice gradually forms a dam, blocking the water behind it from running off. While roofing shingles are made to deal with cascading water, they are not meant to deal with a lot of water backup. The water seeps under the shingles and then under the building paper. It soon soaks the wood or plywood decking and starts wending its way down into the house. The result is water damage, little or large, ruining DRYWALL, PAINT, and lots of other items.

Inadequate insulation is one of the root reasons for this sort of damage. Heat from the house seeps upward, warming the roof and making the snow melt quickly. Coupled with this, there is likely little or no ventilation to keep the underside of the roof cool and allow the snow to melt gradually and evenly.

A wide variety of solutions have been tried to solve ice dam problems: everything from heating cables to slippery bands of metal installed on the lower edges of the roof. Nothing really works, builders say, except insulating and ventilating properly.

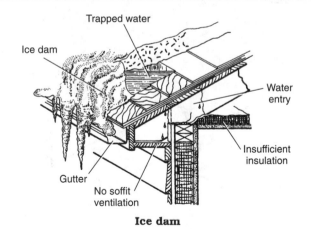

Ice dam

jack rafter Rafter that spans the distance from the wall plate to a hip, or from a valley to a ridge.

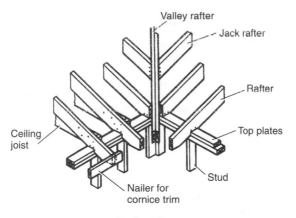

Jack rafter

jack stud Short stud that does not extend from floor to ceiling; for example, a stud that extends from the floor to a window.

jalousie window Window with movable glass slats.

jamb The vertical sides of a DOOR, window, or other opening. The word derives from the French word for leg. In a door or CASEMENT WINDOW, the jamb is broken down into hinge jamb and lock jamb, depending on its location.

jamb block Special block formed with a slot for holding the jamb of window or door frames.

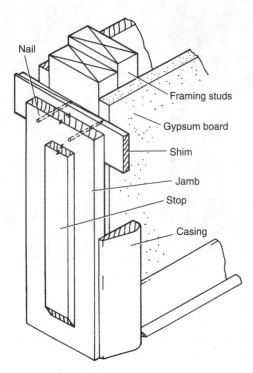

Nail

Framing studs

Gypsum board

Shim

Jamb

Stop

Casing

Jamb block

joint In general, the point where adjacent surfaces meet. *See* MORTAR JOINT.

joint compound The soft material used to fill gaps between DRYWALL boards.

Joint compound comes in powder form as well as ready-mixed. The ready-mixed is far easier to use, and one of the best bargains in building products there is. A five-gallon bucket that contains about 64 pounds of compound can be bought for under $10.

While specifically designed for spackling joints between DRYWALL boards, joint compound can be used for patching and repairing as well. It sticks tenaciously, allowing one to apply it in very thin coats. As such, it is a good material for resurfacing a wall (*see* SKIM COAT). It is not, however, a one-coat material, because it shrinks as it dries. Applying multiple coats brings a surface to the level one needs and can result in a very smooth surface. *See illustration on following page.*

joint reinforcement Steel reinforcement placed in a horizontal MORTAR JOINT.

joist One of a series of parallel beams, usually two inches in thickness, used to support floor and ceiling loads, and supported in turn by larger beams, girders, or bearing walls. *See illustration on following page.*

junction box A type of electrical BOX.

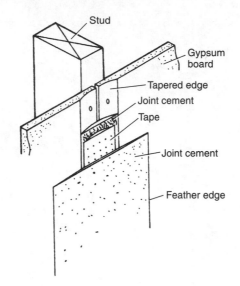

Stud

Gypsum board

Tapered edge

Joint cement

Tape

Joint cement

Feather edge

Joint compound

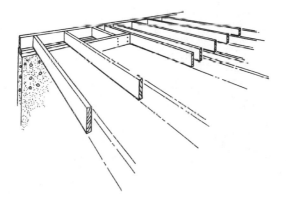

Joists

kerf A saw cut.

keys PLASTER that has seeped through lath and hardened, forming keys that hold the plaster in place.

kick plate Protective metal plate on the bottom of a door.

kiln A furnace or other area used to heat or dry out various materials such as brick and lumber.

kiln-dried Lumber that has been dried out in a kiln.

A tree has a series of tubes or veins that run through it, carrying half the tree's weight in water. When wood goes to the mill, it is cut into boards and allowed to dry, in either the air or a kiln, an oven that reduces its moisture content. If the wood is used as is or dried insufficiently, it will be become a perfect environment for fungi called DRY ROT. Wood like this can twist and warp to such a degree that it can loosen fasteners, with nasty effects on the structure.

Kiln-drying kills fungi and reduces moisture to acceptable levels. The grade stamp carries the story. Any lumber stamped KD-19 means that it was kiln-dried to a moisture level less than 19 percent. KD-15 means it was dried to a moisture content of 15 percent.

kill 1. reduce the suction of materials when plastering by wetting a surface before applying fresh plaster. **2.** In general, to alleviate or eliminate something in building, such as "killing" the shine before painting by scuff sanding.

king closer Brick cut diagonally so it has one cut and one full width end.

king post Vertical wood piece or metal rod that extends from the middle of a tie beam up to the apex of two jointed rafters, creating a TRUSS.

kitchen cabinet Cabinet used in the kitchen.

Kitchen cabinets are made of wood, engineered wood such as PARTI-CLEBOARD covered with PLASTIC LAMINATE, and metal. They are also available in stock, semi-custom and custom sizes.

Stock cabinets come in a series of set sizes. "Base Cabinets" and "hangers," the wall cabinets, are generally available in widths of 12"–48" in 6" increments (12", 18", 24", etc.). Base cabinets are 34½"–35" high; and hangers, 30"–34"—sizes that will fill most of a wall—as well as smaller units (15", 18", 20", 24") to fill the spaces above refrigerators and the like.

Stock cabinets are cheaper than other types of cabinets but can have a distinct disadvantage: They may not completely fill the space available. For example, there might be a wall space of 8 feet 2", but the stock sizes, when set side by side, would only fill 7 feet 10". The cabinets would have to spread out a bit and the gaps covered with filler boards—and that four inches would be lost.

Better for filling space are semi-custom cabinets. Here, manufacturers carry a wide variety of stock parts. An order is given and because the parts available allow a much wider variety of sizes to be built, the chances are they can be made to whatever the exact space needs are.

Custom cabinets are built to fill space exactly according to measurements. The big advantage here, of course, is that no space will be wasted.

Quality of kitchen cabinets vary. Often the best way to determine quality is to examine a wide variety of cabinets in various showrooms and other outlets. After a while, the various qualities will reveal themselves. There are definite quality indicators:

- Good quality wood cabinets will have substantial bottoms for hanging things on, not just something thin and flimsy.
- Front-frame parts, if any, will be assembled with good joinery—TONGUE-AND-GROOVE, RABBETS, or DOWELED together.
- Bottoms should be substantial for hanging things on.
- Doors should be plywood or solid wood.
- Sides should be plywood; individual boards can warp.

The main indication of a good quality PLASTIC LAMINATE-covered cabinet is that the front has a seamless piece of laminate covering it rather than parts. The individual parts of the cabinets have been made, covered with laminate and assembled. If this is the case, you can usually see clips inside the cabinets holding the parts together.

Metal cabinets come only in stock sizes. The finish is sprayed enamel, usually white. Metal cabinets generally warp and bend easily; and if the enamel chips, the metal beneath will show.

See also CABINETS *and* FACE-FRAMED CABINETS.

kitchen triangle The imaginary triangle kitchen designers establish to maximize the efficiency of a kitchen. The triangle extends from the sink to the refrigerator and the stove. The goal of kitchen designers is to design the area to reduce the distance the homemaker must travel between these stations.

knee wall A short wall extending from the floor to the roof in the second story of a 1½ story house. Such walls are knee-high, hence the name. An attic typically has knee walls.

knockout Removable metal disc in an electrical BOX. Such discs may be pried out with a screwdriver to allow cable to be fed through.

kraft paper A heavy brown paper with a variety of building uses. Perhaps its most prominent role is in the installation of plastic laminate. The substrate is coated with CONTACT CEMENT, and the kraft paper is laid on top of it. The laminate is coated and set on the kraft paper. When the contact cement has dried, the kraft paper is slipped out, and the substrate and laminate make contact and bond.

lacing Interweaving roof shingles at intersections.

lag screw This looks like a thick wood screw with a hexagonal top; it is used for a variety of hefty fastening jobs.

Lag screws, also known as *lag bolts*, come CADMIUM-PLATED and GALVANIZED and in various lengths from 1″ to 6″.

Lag screws are used for fastening wood when regular screws are not strong enough. They are usually used in the smaller sizes because they have to be turned with a wrench, pilot holes are drilled for lag screws, and use of an inspect wrench makes them easier to drive. *See also* SCREWS.

Lag screw

lally column Metal column used to support great weight.

lap siding General name for any siding that overlaps, such as CLAPBOARD.

lateral support Sideways support given to a MASONRY wall by bracing, pilasters, cross walls, roofing, etc.

lath The base to which PLASTER is applied. There have been various kinds down through the years, but all work essentially the same way. Each has some sort of openings, and when the wet PLASTER is applied, some of it oozes through the openings and slumps down. When it hardens, these drippings, called KEYS, hold the plaster tight to the wall.

Lath—from colonial times until now

In Colonial times, the first lath used was green oak, and was known as "accordion" lath. A strip of oak was split alternate sides by a froe, and pulled open like an accordion to form a series of Ms, or Ws. This was nailed with hand-wrought nails to the studs and when all the laths were in place, the plaster was applied.

Accordion lath was one day replaced by strip lath, thin strips of wood secured horizontally a small distance apart with machine-cut nails. There were spaces between the strips, so the plaster could ooze behind them and form the keys.

Wood lath was popular well into the 20th century, when it was supplanted by thin sheets of metal with diamond-shaped holes. The plaster would ooze through these holes in the metal lath to form the keys. Metal was replaced, though not completely, by gypsum panels with holes punched in them. These panels continue in popular use today.

lathing 1. Various kinds of LATH used in making PLASTER walls. **2.** The act of applying this material.

lavatory 1. The bathroom itself. **2.** The bathroom sink.

The word comes from the Latin word meaning to wash. There is a variety of styles, colors, and qualities. Most lavatories are made of enameled steel, VITREOUS CHINA, cast polymer, or solid surface materials such as Corian.

Enameled steel is formed steel with a porcelain coating. This is the lowest quality of lavatory available. It is subject to chipping, and makes noise when water drives against it. Perhaps the only thing that can be said in its favor is that it is light.

Vitreous china is hard-burned clay, the same material used to make toilets (see WATER CLOSET). Quality can range from good to not so good, depending on the china used. A sign of a good vireous china is a smooth surface that is free of bubbles, pinholes, or discoloration.

Cast polymer includes a variety of materials that use a polyester resin as a base. The base is mixed with granite, marble, hydrated aluminum, and other materials and then given a gel coating for color and design. Cast polymer imitates other materials, but it also comes in solid colors. It is difficult to tell quality of cast polymer products. One indication is being approved by the Cultured Marble Institute, which monitors products. At times, the gel coating may be too thin, a problem that cannot be repaired easily. Gel coats range from 0.0016 to 0.0020 with the thicker material better.

Enameled cast iron is one of the best lavatories one can buy. Cast iron is tough and inflexible, preventing denting. It is also rare for an enameled cast iron lavatory to chip.

Solid-surface "lavs" are of solid-surface materials and they incorporate all the assets of those materials.

Lavatories come in many different shapes and are installed a number of different ways:

- **Wall hung.** This is hung on the wall, secured to wall studs.
- **Pedestal.** This is a one-piece unit that stands on its own. It has no counter space.
- **Rimmed.** A rimmed lavatory is mounted in a hole made in the countertop.
- **Self-rimming.** This type, also called rimless, has a lip that overlaps the cutout hole in the vanity top.

- **Integral.** This is a unit where lavatory and vanity top come as one piece that is dropped into the vanity cabinet. If this unit is damaged, the whole thing must be replaced. Most integral lavatories are made of cast polymer or a solid-surface material. If these get scratched, of course, repair is possible.

  See also FAUCET.

laying the leads Building the corners of a BRICK wall. Laying the leads properly is extremely important in building a wall because the FIELD brick depend on the corners for being straight, PLUMB, and level.

lay out the bond Calculate how many full BRICK will be in a wall. This is done so the wall can be built with a minimum of cut brick. The foreman on the job will lay out loose brick along the area they are to occupy, including the mortar joints so he knows exactly where he stands.

lay up 1. To place materials in the relative position they will have in the finished building. **2.** Step in making structural wood panels in which veneers or reconstituted wood layers are "stacked" before being pressed into final panels.

L bolts *See* ANCHOR BOLT.

leader Same as DOWNSPOUT. *See* GUTTER.

lead masonry Wall section built up and racked back on successive courses. A line is stretched to leads as a guide for constructing a wall between them.

lead wood A fine mesh steel used to plug holes, particularly when screwholes arc enlarged. When the screw is turned in, it gets enmeshed in and grips the wool, which is soft but compresses into a tight mass.

lean mortar MORTAR that does stick to a trowel. This condition is usually caused by a lack of sand in the mortar mix.

ledger Horizontal board nailed to one surface as a place where other members can rest. A typical use for a ledge is to nail it to the house while building a deck. The ends of the deck members nailed to it.

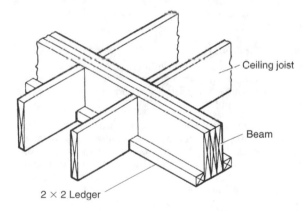

Ceiling joist

Beam

2 × 2 Ledger

Ledger

let-in brace *See* BRACE.

light 1. Space in a window sash for a single pane of glass. **2.** A pane of glass. Windows are sometimes characterized according to the number of lights in them. A window with four panes of glass would be a four light window; one with six, a six-light window, etc.

lightning arrester Device installed to protect house wiring against lightning. Lighting does not need to hit house wires directly to damage them. A lightning stroke near wires can send damaging surges of voltage through them.

Grounding wires properly substantially reduces the risk of damage to wires, but the installation of an arrester reduces the possibility to practically nil. The arrester, which is about the size of a small orange, consists of a housing and three leads. Hooking it up is quite simple.

lightweight aggregate *See* AGGREGATE.

lime A caustic substance created by the action of heat on limestone, shells, or other forms of calcium carbonate. The heat drives carbonic acid out, leaving only quicklime, an important ingredient in PLASTER and MORTAR, particularly for its plasticity.

limestone A natural stone that can be used as itself as a building material or as a source of lime.

line pin In masonry, a metal pin used to attach a string line used for aligning masonry units.

lineal foot Running foot, literally in line.

linseed oil Oil made from flax plants.

Until ALKYDS came along, linseed oil was the main ingredient of most PAINT. It is still an important ingredient in some paints and many stains. Its most important characteristic is that it hardens when exposed to oxygen. (It was also used in the making of linoleum, for which it was named.)

Linseed oil is sometimes used pure—after dilution with mineral spirits or another solvent—on furniture, rubbing it deeply into the wood. On the other hand, a pure use is not recommended for exterior surfaces, where it can be considered food for mildew. In fact, mildew are not the only form of life to like it. Cows like it, too. Linseed oil is one of the more dangerous products the contractor can use because its flammability is hidden.

Cows like it, too

Cows have been known to lick surfaces freshly painted with linseed-oil-based paint, simply because they like it just as they like the flax plants it comes from. This can make the cows sick or bring on an early demise, but they do it.

Spontaneous combustion

Rags or pads used in applying linseed oil must be carefully disposed of after using them or they can catch fire by spontaneous combustion. There have been quite a few fatalities and injuries for people who did not know this could happen.

---

lintel Horizontal wood or MASONRY structural member that supports the load over an opening such as a door or window. The word derives from the classical Latin word "linen," which originally meant THRESHOLD. It has gradually changed in meaning over the years to describe what it does now.

lintel block U- or W-shaped CONCRETE BLOCK used in construction or horizontal BOND BEAMS and LINTELS at the job site.

listed Electrical product that has been tested and found sound by the UNDERWRITERS LABORATORY.

The Underwriters Laboratories explains that being listed does not mean an electrical device has multiple purposes. It means only that when tested to perform the specific task it was designed for, it passed. Nor does it mean that the device is of high quality. It may not be. For example, if when testing a switch the UL flipped it 5000 times and it didn't malfunction until the five thousand first time, it may pass because the UL just demanded that it be able to be flipped 5000 times without malfunction. Another switch performing the same task might be flipped 25,000 times without a problem. The second is of higher quality, but both are listed, and the UL will not say which is of higher quality.

live load The load, expressed in pounds per square inch, of people, furniture, snow, etc., that are in addition to the weight.

Live load is another factor engineers figure when calculating how strong framing needs to be. Building codes set various figures for weights that framing must support. Normally, the figures represent weights that are far in excess of what day-to-day loads will be.

load-bearing walls Walls that support the weight of a structure as well as themselves. *See also* PARTITION WALL.

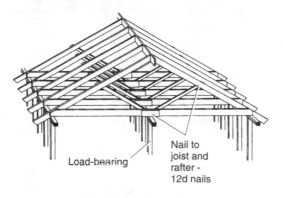

Load-bearing

Nail to joist and rafter - 12d nails

Load-bearing wall

load center The distribution center for electrical power within a house. Also called PANEL BOX.

loads 1. The pressure or weight a building must sustain.

Loads are of various types and are, of course, a key consideration in planning how to erect a building. Concentrated loads are applied over a small area such as a spiked heel, a shoe, or the leg of a piano. "Dead" loads are stationary, permanent loads—the weight of the material used to erect the building. "Live" loads are those the structure must be able to support under normal conditions, such as furniture, people, or equipment that would be placed on a surface. Such loads are normally specified by the BUILDING CODE. Uniform loads are those that are evenly distributed over a large area. Live loads and uniform loads are sometimes considered the same.

2. In electrical current, the current drawn by an appliance or other device when it is turned on.

long-and-short work In MASONRY, a method of installing stone in window and door frames.

lookout A short wood bracket or cantilever that supports an overhang portion of a roof, usually concealed from view by a soffit, which is nailed to it. Also, because of its shape, known as a "porchchop." *See illustration on following page.*

loose fill *See* INSULATION.

louver An opening with a series of horizontal slats arranged to permit ventilation but to exclude rain, sunlight, or viewing into. The word originally was used to describe a stone turret in medieval times.

low-e glass Glass impregnated with special metals to make it insulating.

low voltage Voltage less than twelve volts. Low voltage is not hazardous and is used for a variety of WIRES. Bell wire, for example, is commonly known as low-voltage wire.

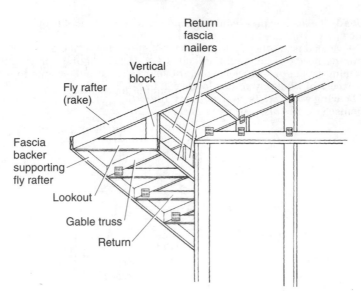

Return
fascia
nailers

Vertical
block

Fly rafter
(rake)

Fascia
backer
supporting
fly rafter

Lookout

Gable truss

Return

Lookout

lug sill Window sill that extends beyond the width of the window and whose ends are recessed in the wall.

lumber General term for wood used in building.

In the wood-products industry, lumber is broken down into a number of product categories: timbers, DIMENSION LUMBER, and BOARDS, as well as SOFTWOOD and HARDWOOD. Lumber is graded according to a number of characteristics.

Softwood lumber is divided into three size categories: finish grades nominally 1″ thick are called boards. Lumber with a thickness from 2″–4″ in dimension lumber. Lumber 5″ and thicker is called 2 by. Most framing materials in light-frame construction are dimension lumber.

Dimension lumber is sold by grade, species (or subspecies group), and size. Lumber sizes are given as the NOMINAL dimensions. At the time a log is first sawn from a tree, it may approach those dimensions, but resawing, seasoning, and surfacing reduce it considerably.

Dimension lumber comes in 2′ increments. Most common sizes are 8′, 10′, 12′, 14′, and 16′, but most lumberyards stock longer lengths. Actual lengths are usually slightly longer than nominal size to allow for trimming. Most framing lumber is surfaced on four sides (S4S) and has eased edges (EE), although some square stock is manufactured.

To assign lumber grades, a certified grader evaluates both natural and manufacturing characteristics. Unlike boards, framing lumber is graded primarily for strength rather than appearance. Lumber grading sets quality control standards among lumber mills manufacturing the same for similar products. All lumber is stamped with its particular grade.

All major BUILDING CODES require that lumber used structurally contain five basic elements: the symbol or lumber logo of the quality control agency; the mill number or name; the grade of the material; the species or species combination; and the condition of seasoning at the time of manufacture.

Lumber making—then and now

In Colonial times, there were no lumberyards, but builders had a ready source of wood—the vast forests of America. All they had to do was cut down what they needed and process it with the techniques they had developed.

One of the chief techniques in making lumber was splitting. Trees tend to split along their grain line, so if one wanted to make narrow but long lengths from a log, one would typically use mallets and wedges to split it along the grain, just as firewood is split along the grain today.

Big beams were made by a method known as hewing. The builder would find a straight tree and mark one edge of the beam he wanted along it using chalk or beetle juice. Then, standing on the log, he would use a long ax to make a series of fairly close-together cuts in the side of the log, cutting just deep enough that the cut bottomed at the line (this method was known as "hewing the line.") He would chop out the pieces between the cuts with a broad ax and repeat the procedure on the other sides.

While some lumber was sawn, splitting was the preferred method for most lumber-making. Then, in 1624, the first sawmill was built in Jamestown, New York. This relied on primitive methods, but in 1814 the circular saw was brought to use by Benjamin Cummings, a blacksmith in Bentonville, New York.

Fifty years later, the band saw was found to be superior to the circular saw. It could cut larger logs and did not chew up as much wood as the circular saw when cutting. Today, the band saw is still at the heart of lumber making.

lumen The amount of light that falls on one square foot of area from a distance of 1'.

LVL Laminated veneer lumber, material made by parallel lamination of veneers into thicknesses common to solid sawn lumber (3/4" to 2 1/4"). It is considered engineered LUMBER.

LVL was first developed in the 1940s, when it was considered for building high strength aircraft. Research gradually produced a material that utilizes veneers 1/8"–1/10" thick, hot-pressed together with phenol-formaldehyde adhesives into lengths from 8'–60' or more. Various joints are used in assembling the material (butt and scarf); but for sections up to 8' long, no end joints are required, and SCARF or finger JOINTS can be used to join these lengths to one another or to sawn lumber. LVL differs from PLYWOOD in that plywood veneers are placed crisscross to one another; LVL veneers are placed in the same direction.

I BEAMS are also made using laminated veneer lumber. Here, there are flanges flanking panel-style webs made of hardboard and flakeboard.

There are many uses for laminated veneer lumber, which is known for its strength. In home construction, LVL is ordinarily used as joists on upper floors of large houses (those in the 5000 to 6000 square foot range), as well as headers in various situations.

LVL is not meant, however, to be in the open because such boards are not good looking. *See illustrations on following page.*

Harvesting lumber

Big lumber companies seem to have succeeded mightily not only in conservation efforts in forests, but in wood products that are even better than Mother Nature has to offer. Today, such companies have their own forests, where they raise trees that can reach maturity in eight to twenty years, and then process them in ways that can yield material that is stronger than lumber processed from "old growth forests."

LVL

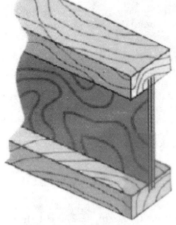

LVL

Traditional
lumber

LVL: Cutaway view

macadam A paving material made of compacted small stone. Macadam was named after its Scottish inventor, John McAdam, who combined it with tar and called it TARMAC.

madison clips Thin metal clips for attaching switches and receptacles in OLD WORK installations. Commonly known as BATTLESHIPS.

male In plumbing or electrical products, the part of the product that is inserted into a FEMALE part.

mantel The "frame" around a fireplace, including the shelf which is ordinarily connected to the breast of the CHIMNEY above the fireplace.

marble A natural stone that comes in a wide variety of colors.

Marble is essentially recrystallized limestone. After earthen materials are crystallized into limestone, it is subjected to heat and pressure that recrystallize it into a form much harder than the original. Innumerable materials work their way into limestone, which is why it is available with such widely varying colors and patterns, ranging from white-and-gray mottling to black. MARBLE may have directional grain or be more generalized.

Although marble can withstand heavy use, it can be stained and abraded. Color can also affect durability. Colored marble is usually denser than white marble, which stains more easily and more apparently. Dark marble is usually harder and more compact and less likely to stain. However, because the surface is dark, scratches show more.

Marble is available in a variety of forms, including tile and solid surface material.

masonry Any material made of concrete or stone, including beck, block, poured concrete, stone—anything that's hard and stone-like in its properties. Many people draw a distinction between masonry and concrete, defining the former as anything composed of separate entities that are assembled—such as a brick and block—and viewing concrete as a mass that is poured, rather than masonry.

147

masonry cement Manufactured mix of PORTLAND CEMENT, LIME, and other materials.

mastic Another name for CONSTRUCTION ADHESIVE. Characteristically it is dispensed from a caulking gun with sort of clay-like consistency.

matched boards Boards with edges tooled to fit together well, such as TONGUE-AND-GROOVE.

MDF Medium density fiberboard. Like other FIBERBOARD, made with wood particles bonded together and compressed into sheets. MDF is a useful material in making cabinets and the like because it holds paint well and does not shrink and expand to the same degree as real wood.

mensuration Measuring, particular as to length, area, and volume in building.

mesh tape Self-adhesive fiberglass tape. It can be used for hanging new DRYWALL as well as in repairs. Studies by one major drywall company indicate that mesh tape does not hold up as well as the more usual paper tape.

metal ties General term for metal strips that connect a variety of items. For example, metal ties are used to hold together CAVITY WALLS.

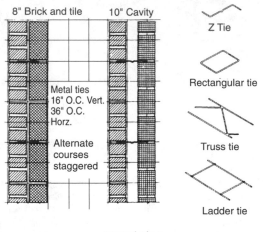

Metal ties

mildew Stain fungi or microscopic plant life.

Mildew is the most common cause of house paint and unfinished wood discoloration. While most mildew spores are black, it can be red, green, and other colors. It grows extensively in warm, humid climates, but it is also found in cold northern states.

Mildew may be found anywhere on a building, but it is most commonly found on walls, behind trees or shrubs, or where air movement is restricted. Mildew may also be found where there are certain dew patterns on the house. Dew forms on those parts of a house that are not heated and cool rapidly, such as eaves and ceilings of carports and porches. This dew provides a source of moisture, one of the three things mildew likes best, the others being dirt and darkness.

One way to distinguish mildew from dirt is to examine it under a high-power magnifying glass. In the growing stage, when the surface is damp or wet, the fungus is characterized by its threadlike growth. In the dormant stage, when the surface is dry, it has numerous egg-shaped spores; by contrast, granular particles of dirt appear irregular in size and shape.

The presence of mildew can be checked by applying a drop or two of household bleach (five percent sodium hypochlorite) to the stain. The dark color of mildew will usually bleach out in a minute or two. If the stain does not respond to the bleach, it is probably dirt. (Use fresh bleach; it deteriorates and loses its potency when it stands around.)

Before painting or repainting, all mildew must be killed or it will grow through the new paint. The standard solution used for killing mildew is ⅓ cup of household detergent and one quart of bleach to 3 quarts of water. The detergent must be ammonia-free: mixed together, the two produce a deadly gas.

The solution should be applied with a stiff brush and followed up with fresh water. The bleach solution can also be used inside the house, such as in the bath (where mildew commonly forms) in the same way it is used outside. A good preventive tip is to add mildewcide to the paint. If a stain or clear is used, it should also contain a mildewcide.

There is also a wide variety of commercial products available for removing mildew, some of which are just sprayed on and hosed off.

mildewcide Chemical added to coatings to retard the growth of mildew. Mildewcides are poisons that become part of the surface film that paint creates. Mildew does not form because it is killed when it comes in contact with the material.

mill glaze Hard, shiny spots on wood created by the binding action of saws when wood is being cut at the mill.

millwork Wood that has been partly finished at the mill.

milk 1. Lingo for an acrylic additive used with floor leveling compounds. When laying tile and sheet goods, flooring often involves leveling the floor with a compound of some sort. Thin plaster-like compounds are often used, but these are likely to crack and release from the floor when spread thinly over a wide area. To make it strong and more flexible, the acrylic milk is used to mix the material instead of water. **2.** Additive for ceramic tile.

mineral granules Natural or synthetic stones in various colors. These granules provide a wearing surface for ROOFING.

mineral-surfaced roofing FELT coated with ASPHALT and MINERAL GRANULES and produced in roll form.

mineral wool A fluffy type of INSULATION. Until fiberglass insulation came along, mineral wool was perhaps the most common material used to insulate walls and ceilings. Today, it is not used nearly as much. Mineral wool is made by subjecting rock or molten slag to high-pressure steam blasts; a number of variations are available.

miter joint Joint made by cutting material ends at 45 degrees. Miter joints are commonly used on MOLDING. Miter joints are used universally to assemble molding used around doors and windows.

Miter joint

modular A unit of the same size used to build something. Modular construction is easier than the non-modular type because dealing with materials that are the same size is easier than dealing with different sizes. *See* MODULAR MASONRY UNIT.

modular masonry unit Concrete block with NOMINAL dimensions equal to
the size of the masonry unit as made in the factory, plus the thickness of
one mortar joint.

moisture barrier Any material that will block the flow of moisture through
space. A common moisture barrier is polyethylene sheeting, which is used
over unfaced insulation to block the flow of moisture-laden heat and
through it and as a ground cover to keep moisture from transferring to a
SLAB or CRAWL SPACE FOUNDATION.

molding Wood trim pieces.

For good looks, molding is absolutely essential. It gives doors, windows,
and rooms that finished look. Molding is generally available in hardwoods
such as oak when it is designed to be clear-coated, and in pine when it is
meant to be painted.

Case molding, also called casing, is the flat molding used to form the
edge trim around interior door openings as well as the room side of exterior
door frames. Case molding varies in width, usually being between 2¼" and
3½", depending on style; thickness varies from ½" to ¾", although ¹¹⁄₁₆" is
standard in many narrower moldings. Casing around the window frames is
or should be the same as around the door frames.

Which joints are used in installing casing depends on whether it is
shaped or molded. Casing with a molded shape must have a mitered joint
at the corners, while square-edged material can have a butt joint. Moisture
content of the molding also matters. If the moisture is above the recommended limit then a miter joint might open as the wood dries out. Carpenters minimize this by installing gluing on a small "spline" at the corner of
the mitered joint. Using a spline is recommended wherever moisture may
be a problem.

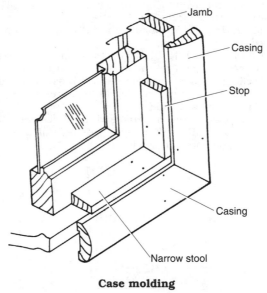

Case molding

The sash stops (the molding the double hung windows rides up and
down in), the "stool," and the apron are all considered molding. Sometimes
case molding is used like a picture frame around a window opening, and
the stool becomes a filler member between the bottom sash rail and the
bottom casing.

Base molding, also known as "shoe" and "toe" molding, serves as finish pieces between the finished wall and the floor. It is available in several widths and forms. Two-piece bases consist of a baseboard topped with a small base cap. When the wall finish is not straight and true, the small base molding will conform more closely to the variations than the wider base alone will. A common size for this base is $5/8'' \times 3\frac{1}{4}''$. One-piece base moldings vary in size from $7/16'' \times 2\frac{1}{4}''$ to $\frac{1}{2}'' \times 3\frac{1}{4}''$ and wider. Common molding styles used are *traditional* and *ranch*.

Ceiling moldings are installed at the junction of wall and ceiling. As with base moldings, the inside corners should be COPE molded. This insures a tight joint and will retain a good fit if there are slight moisture changes. Molding styles include *crown* and *cove*.

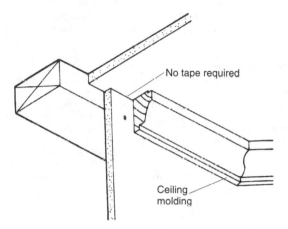

No tape required

Ceiling molding

Ceiling molding

mole run Meandering ridge in a roofing membrane.

molly Brand name of a common hollow wall fastener.

Mollies come in a variety of sizes for different weight objects and are used to hang objects from DRYWALL. A hole is drilled in the material, the fastener slipped into the hole, and the screw on it tightened. As this occurs, the device's prong-like projections expand inside the wall and grip its back side. The screw is withdrawn, then slipped through hole in object.

monkey s--t Slang for electrician's putty, also known as *dum dum*.

mopping Applying hot BITUMEN by hand or with a mechanical applicator. Mopping is done in a variety of ways, from applying it in roughly circular areas to strips.

mortar There are three kinds of mortar that can be used. The first of these is cement mortar, which is simply PORTLAND CEMENT mixed with water. The only difference between this mortar and concrete is that concrete contains AGGREGATE. As such, the mortar is very strong, but it is not very workable, being stiff, and it cannot be mixed with enough water to make it plastic without reducing its strength.

Cement mortar is used for foundation walls, particularly ones that may have to endure hydrostatic pressure.

Cement-lime mortar is used because it combines strength with workability, depending on the proportion of lime to cement—as the lime increases, so does the workability as long as the amount of sand used is the same. There are four types of mortar used.

Type M is the strongest of the cement-lime mortar mixes made. In addition to sand, it is composed of one part Portland cement to one quarter-part lime. Type M mortar is used when the weather is bad. It can also be in contact with soil, and may be used in reinforced concrete and retaining walls.

Type N, stronger than Type O, contains one part lime to one part Portland cement. Type N mortar is a fairly strong mortar that can be used when the temperature goes below freezing. It can be found in chimneys, exterior walls, and other structures.

Type O mortar contains two parts lime to every one part cement. The weakest of the mortars used, Type O mortar is used for interior brickwork or where the bricks are not exposed to freezing temperatures.

Type S mortar is also a medium-strength mortar that is particularly useful where a wall is subjected to sideways or lateral stress—being pulled on. As such, it is a good mortar for use in ceramic tile MUD jobs. To make Type S mortar, one part Poland cement is combined with sand and one half-part lime.

Ancient mortar

Thousands of years ago the ancients, chiefly Roman and Greeks, used mortar made completely of lime. They would crush calcium and magnesium carbonate or calcium carbonate, and heat them in a oven to rid the lime of its water. The resulting product was called quicklime. This was slaked (water was added to it) until it became a slurry.

This slurry, called lime putty, was stored for awhile—sometimes quite awhile. The longer it was not used, the more workable it became. When it was finally used, it was quite a good mortar. Indeed, structures built 2000 years ago are still standing.

See also HYDRATED LIME.

mortar joint Joint made with MORTAR to connect masonry units, such as BRICK and BLOCK.

Mortar is a key element in solid MASONRY construction. The better the joints are, the more durable and weatherproof the structure is.

The key characteristic of a well-made mortar joint is the lack of gaps or cavities. Gaps or cavities mean weakness, breakdown, and the loss of weathertightness, which can lead to further breakdown.

In general, a mason will ensure a good joint by first laying a generous bed of mortar, furrowing it only slightly, and applying a good amount of mortar to the end of the brick. When the brick is tapped down, the excess mortar squeezes out and is clipped off neatly. It is normally good practice to set the brick in place within one minute of laying the mortar.

A number of mortar joints are extant, as follows:

- **Bed.** This is a horizontal joint. A layer of mortar is placed and furrowed slightly with the tip of the trowel. The brick is placed, and any excess mortar is trimmed.
- **Collar.** The vertical joint between wythes of brick. It is important not to "slush" such joints, because it often results in incompletely filling the joint.
- **Head.** The joint between the vertical ends of masonry units.

mortar joint finish The look given to a mortar joint.

In general, there are two types of mortar joint finishes, the so-called *troweled* and *tooled*. The former may be done with the trowel; excess mortar is just "struck" off and that's that. A tooled joint is achieved by pressing a "jointer" tool along the wet mortar, compressing and shaping it. Following is a lineup of joints. Each type of joint has an overall purpose.

Concave and V-shaped joints are accomplished with a jointer. Good for resisting rain, they are recommended for areas subject to heavy winds and rain.

A weathered joint is the best of the troweled joints. It is compact and sheds water easily. Making the joint requires care because it must be worked from below.

Very common, the stuck joint leaves a ledge where water can collect, making for a less watertight joint than concave, V-shaped, or weathered joints. As American masons work from inside the wall, the stuck joint is easy to strike with a trowel.

Rough-cut, or flush, joints are made by laying the trowel against the brick and slicing the mortar off in any direction. As such, it is the easiest kind of joint for the mason to make, but produces an uncompacted joint with a hairline crack where the mortar is pulled away from the brick by the cutting action. Such joints are not always waterproof.

Raked joints are made by cutting out a portion of the mortar, leaving a fairly substantial depression in the joint. As such, a raked joint is not particularly weatherproof and should not be used in areas with substantial rainfall. Raked joints produce noticeable shadows and darken the overall look of a wall.

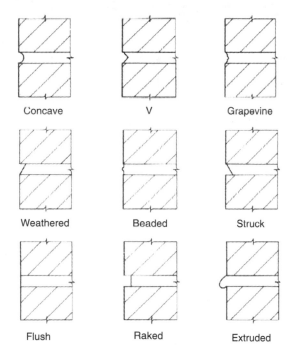

Concave	V	Grapevine
Weathered	Beaded	Struck
Flush	Raked	Extruded

Various types of mortar joints

mortise and tenon Slot cut into a board, plank, or timber, usually edgewise. The mortise receives a projecting part, or tenon, of another board, plank, or timber to form a joint.

Many years ago, mortise and tenon joints were used in house building, a tedious and exacting process (*see* BRACED FRAME). Today they are mainly used in assembling cabinets. *See illustration on following page.*

mud 1. JOINT COMPOUND **2.** Cement used to set CERAMIC TILE.

mullion Vertical member of a WINDOW dividing the lights or window panes. A window with such vertical parts is sometimes known as a mullion window.

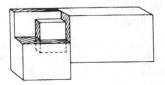

Mortise and tenon joint

muntin A short bar, horizontal or vertical, separating panes of glass in a WINDOW sash.

muriatic acid Material used to clean masonry.

Muriatic acid is the most common cleaner of BRICK. It is actually sulfuric acid mixed with water for use, the dilution depending on the job at hand. For safety, the acid is always poured into the water to minimize the acid's splashing back at the user.

nail Pointed metal rod with a head, used for fastening a wide variety of materials.

Nails are generally available in lengths that range from 1″ to 6″; as the nail lengthens, the diameter thickens correspondingly. Nails are commonly sized by weight and number. Weight is expressed by the letter "d," which represents pennyweight, the way nails used to be sold, with sizes running from 2d (1″ long by ⅛″ thick) to 60d (6″ long by ⅜″ thick).

Some nails can be used for a variety of purposes while others, such as the roofing nail, have a specific purpose. Following is a lineup of nails.

A box nail is a thin, rosin-coated nail, commonly used when the wood is in danger of splitting. The box nail is commonly used for wood flooring and subflooring, working particularly well in securing oak and pine. The rosin on the nails is the key to their sticking. As the nail is driven in the rosin heats up and when it cools the nail grips tenaciously.

Carpenters take particular care when using box nails. Thinner than the common nail, box nails bend more easily.

Common nails, plain shank nails with a large round head, are the most commonly used nail for construction, particularly framing. They are available plain or galvanized; the latter has a rough surface, gripping better and resisting rust. They also cost more. *See illustrations on following pages.*

Casing nails are slightly thicker than the finishing nail used to secure trim. Casing nails have a flat head rather than a cupped one, like a finishing nail. They are normally driven into the surface flush and painted over.

Finishing nails are slender, with small, cupped heads. While available in various lengths, most finishing nails used in residential construction are 1½″ long. These nails are designed to be used where the nailheads are to be hidden. The small heads may be left flush with the surface, but usually they are "countersunk," their heads driven slightly beneath the surface

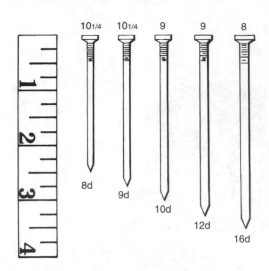

Common nails

and the holes filled with wood putty and sanded so that the nail is completely hidden.

Finishing nail

Some carpenters working on expensive wood will use a tool called a "magnetic spinner" to lessen the chance of damaging the wood while driving the nail. The tool holds the nail and actually spins it into the wood up to the point that it can be countersunk.

A cut nail is tapered, with a flat top and tip used to secure tongue-and-groove flooring and wood to masonry. (*See* FLOORING, TONGUE-AND-GROOVE.) The only practical way to drive cut nails is with a nailing machine. The machine sets up and drives each nail.

Cut nails are sometimes called "square" nails, but they are different. Square nails are the old hand-wrought type, which truly were square, while cut nails are rectangular.

Cut nail

A double-headed scaffold nail is the same as a common nail with a standard head and neck or rim of metal a little lower than the head. It is used for the temporary assembly of scaffolding on a construction job. The nails (available in various sizes up to 3″ long) are driven in the rim of metal. As such, for disassembly the nails are easier to get a grip on and pull.

Double-headed scaffold nail

The large-headed drywall nail comes in two forms. One is blued and threaded; the other is plain steel. The plain steel type has a smaller head; standard lengths are 1⅛″ and 1¼″.

Drywall nails are meant to be driven in so they DIMPLE the surface (JOINT COMPOUND is used to cover the depressions). In practice, because of the way the heads are designed, the plain steel is better. Blued drywall nails tend to cut the paper surface of the drywall, which can lead to the nails' loosening. Blued drywall nails are also more difficult to finish over.

Drywall nail

A masonry nail looks like a common nail except it is thicker and fluted. Masonry nails are used to secure wood and other materials to masonry. They are driven with a two-pound sledge. The nails normally should penetrate the masonry no more than an inch; indeed, it is unlikely they will go in deeper. If they stall, they will not penetrate further. There is danger of the head being whacked off if efforts continue to drive it in further. See also MASONRY SCREWS.

Small-headed nails used to secure wall paneling are available in various colors to match the paneling. When in place, panel nails are not noticeable. They are also good for securing pre-stained molding.

Roofing nails are large-headed and have a smooth shank used to secure ASPHALT ROOFING SHINGLES and FELTS. They are available in lengths from ¾″–2″ with the size depending on the thickness of the material being driven through (the 1½″ size is the most popular). The roofing nail must penetrate the shingle, felt, and DECK, punching holes and creating splinters. It is these splinters that actually keep the nail from working loose.

Spiral-threaded nails are used to secure underlayment. There are two kinds of spiral nail. One has annular spiral threads and comes in 1″ and ¼″ lengths. The other is fluted and has a spiral head and is longer, 2½″ to 3″.

Spiral nails have tenacious gripping power because they turn and act like a screw when driven. The smaller nail is used to secure hardboard underlayment while the larger is used for WAFERBOARD, which is thicker and therefore requires a longer nail. It is also used for textured hardboard siding. See HARDBOARD, and SIDING.

Spiral nails are economical, costing half the price of screws, and being twice as fast to use.

Nail size

The practice of nails being referred to in terms of penny size goes back to fifteenth-century England, when nails were sold by the pennyweight. Pennyweight was represented by the letter D from the Roman denarius coin, also the name for an English penny. A ten penny nail cost ten pennies, an eight penny nail eight pennies, and so forth.

National Electric Code A set of standards for installing electrical facilities.

The National Electric Code (NEC) is an arm of the National Fire Protection Association, a non-profit organization that promotes safety in electrical installations in the USA.

The code established by this organization is not compulsory, but many communities have adopted it entirely for use as their local electrical code. Others have modified it somewhat, based on local conditions. Still others have used it as a basis for even more stringent code. It should be remembered that the National Electric Code contains minimum safety standards. *See also* BUILDING CODE.

neat cement Pure cement not mixed with a sand admixture.

newel Post to which the end of stair railing or BALUSTRADE is fastened.

nickel-plated Refers to items that have been coated with nickel, which is not quite as shiny as chrome. Such items are more attractive and more durable than CADMIUM-PLATED pieces, although not weatherproof.

nineteen-inch selvage Asphalt roofing with a 17″ granulated surface and 19″ non-granulated. Also called *SIS* or *wide selvage asphalt.*

nipple In plumbing, a short section of pipe used to extend the length of pipe in conjunction with another fitting. *See* FITTINGS.

NMC Non-metallic cable. *See* ROMEX.

nogging Brickwork installed between house timbers. Vertical timbers are called posts; horizontal sections are called "nogging pieces."

nominal size The named size of an item. Named size may not be ACTUAL size. Lumber that is nominally 2″ × 4″ is actually 1¾″ × 3½″. The practice even extends to such things as tarpaulins, where one can purchase a 30′ × 40′ tarpaulin, the dimensions of which will be slightly smaller.

nonbearing partition wall WALL that does not carry any weight.

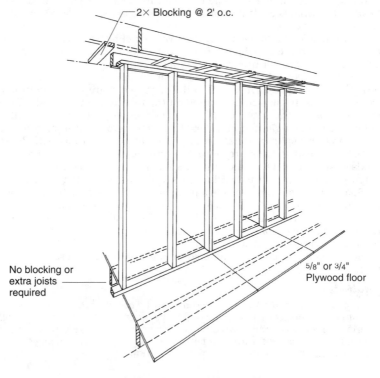

Nonbearing partition wall

nosing The rounded or otherwise shaped edge of something such as a stair TREAD or window STOOL.

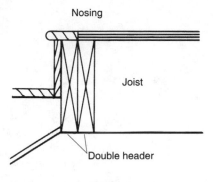

Nosing

notch A crosswise RABBET at the end of a board.

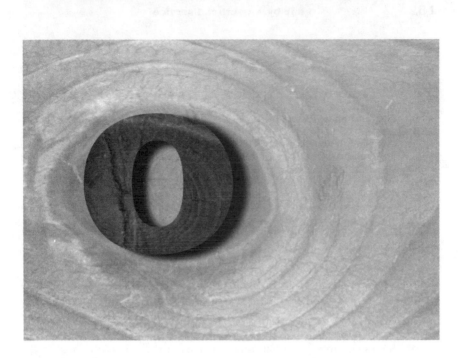

O.C. On center. The measurement of spacing for elements such as STUDS, RAFTERS, and JOISTS from the center of one member to the center of the next.

For most construction, studs and other framing members are placed 16″ apart—16″ O.C. Doing this is not only structurally sound in most cases, but makes it easier for someone working on the house later. For example, when placing a hanging KITCHEN CABINET, it is advisable to screw the backs to studs. Finding one lets the carpenter find others quite easily.

The 16″ spacing also allows one to use panel material, such as PLYWOOD, more easily. Such sheets are commonly 4′ × 8′, 48″ wide by 96″ long. This allows for placing adjacent panels on studs so that the end edges will fall on studs and ensure that there is solid nailing surface throughout.

Some construction allows for stud spacing of 24″. This spacing, also, allows for the panel edges to fall on studs.

Working on an older home can be more difficult than its modern counterpart because many times the member spacing is not consistent. The carpenter of yesterday was more interested in getting a house erected than worrying about an extension that could some day be put on.

offset 1. In plumbing, a combination of FITTINGS that makes two changes in direction, bringing one section of pipe out of line but into a line parallel with the other section. **2.** In masonry or building, a recess or sunken panel of some sort. **3.** A horizontal ledge on a wall created by the narrowing of the wall at that point.

ogee Molding that in profile has a S shape, the result of the joining of concave and convex shapes.

old work In the electrical field, doing electrical work involving existing work.

one by Trade term to refer to any lumber that is nominally 1″ thick . . .1″ × 2″, 1″ × 3″, 1″ × 4″, etc.

open-end block Concrete block with an end removed and used for placing the block around vertical steel reinforcement. *See* BLOCK.

open-string stairs Stairs with one side against a wall and the other side having a HANDRAIL.

oriented strand board *See* OSB.

OSB Oriented strand board. A type of structural flakeboard composed of layers, with each layer consisting of compressed strand-like wood particles in one direction, and with layers oriented at right angles to each other, all bonded with phenolic resin glue. *See also* FIBERBOARD.

out of true Not perfectly perpendicular.

outlet Electrical receptacle

outlet sewer The section of pipe between a SEPTIC TANK and the drainage ield in a septic system.

overcurrent devices Technical name for FUSES and CIRCUIT BREAKERS.

overflow In a tub, lavatory, or sink, the opening through which water will flow if it rises to a certain point. The purpose of the overflow is to prevent the particular fixture from overflowing.

overhand work Laying BRICK from inside a wall by workers standing inside the wall or on a scaffold.

overhead service Term used to describe the delivery of ELECTRIC SERVICE to a building via above-ground wires.

Electric power comes to the typical house by means of overhead service. The cables run from the utility's power lines at the side of the road to the house. The three wires are known as the "service drop," and they connect to the house via insulated fittings. Each of the three wires contains a deep loop, known as a "drip loop" the purpose of which is to keep rain from running into the house, which could happen if the wires ran straight. From there, the electricity runs down a metal conduit called a "riser" and into the house.

When wires run overhead, a variety of restrictions are put on their position, including their height above houses, driveways, sidewalks, or the roof of another building.

The connection of the wires at the house must also be done so that cables cannot do any damage. Their weight exercises a lot of pull on the house, and this could conceivably result in ripping away a portion of the house or even to making the house slightly lopsided. To prevent this, the service drop is often connected to the corner of a house. Classically the strongest point of the house, the corner is reinforced if necessary. If the house is not considered strong enough to resist the pull, a separate "meter pole" is erected between the utility's power line and the house.

If the house is too low to comply with codes regarding how high up the cables must be, a 4″ × 4″ post can be through-bolted to the house and the cable connected there. There is a variety of other ways to accomplish the same goal. *See also* UNDERGROUND SERVICE.

packing Fibrous material used to help make a FAUCET leakproof.

pad stone 1. A LINTEL. **2.** A large stone placed under a BEAM or GIRDER to help support weight.

paint Opaque liquid film used to protect and preserve interior and exterior surfaces.

Paint can be made a great many ways, but it is essentially composed of pigment, vehicle, and additives. The pigment is solid grains or powder particles that give paint color and hiding power, depending on the type and ratio of pigment to vehicle. It can also contribute to such properties as hardness, corrosion resistance, scrubbability, and adhesion.

The vehicle portion has three important parts: thinner, binder, and additives.

The thinner gives paint its liquid appearance in the can and works with the binder and additives to form the paint vehicle. Water is the thinner in latex paints; mineral spirits act as the thinner in oil-based paints. When the paint is applied, the thinner evaporates.

The binder "cements" the pigment particles into a uniform paint film and makes the paint adhere to a surface. The type and quality of binder used also determine performance characteristics of the paint.

Additives are substances added to paint to give it properties the paint will not have by itself, wet or dry. For example, "thickeners" increase the thickness of the paint; "driers" speed the drying process, and "anti-foaming agents" minimize foaming and formation of air bubbles.

See also ENAMEL *and* PAINT FINISHES.

A long history

Paint goes back a long way. The earliest known use of surface coatings goes back as far as 2000 B.C. Early Chinese and Egyptian artisans used mixtures of drying oils, resins, and pigments for pictures and inscriptions

163

on their tombs and temples. Surprisingly, these paints closely resemble—in makeup and appearance—the more fundamental types in use today.

As the population expanded, and people began to travel, trade, and go to war on an organized scale, the need and desire for decorative coatings grew. The ancients used paints on their ships, utensils, musical instruments, weapons, and palaces in an ever-growing variety of pigments and binders. White pigments came from white lead and natural white earths such as clay, gypsum, and whiting. Black pigments were charcoal, lampblack, boneblack, natural graphite, and powdered coal. Among the yellow pigments were ochres, gold powders, and lithage. Reds came from iron oxides, red lead, cinnabar, and natural red dyes. There were a number of blues, such as Egyptian blue, lapis lazuli (ultramarine), copper carbonate, and indigo. Among the greens were terre verte, malachite, verdigris, and natural dyes. Their binding media included gum arabic, glue, eggs, gelatines, beeswax, pitch, shellac, animal fats, and saps of various trees, as well as drying oils.

The amount of paint made was not much by modern standards. A generally low standard of living, scarcity of raw materials, and slow processing of paint by hand resulted in a slow growth in the use of paint.

Man's inventiveness led to developing better manufacturing methods. In 1200 AD, a monk named Presbyster described the making of a varnish based completely on non-volatile constituents, chiefly drying oils. About 1500, the first modern varnishes were made by "running": resin with sandarac in linseed oil. These varnishes were mainly used to decorate and protect crossbows and other weapons.

During the next 300 years, the most popular resin used for both protection and decoration was amber, either alone or in combination with linseed oil. The scarcity of amber led to a search for substitutes, and it was replaced almost completely with fossil and semi-fossil gums such as copal, gum arabon, and gum elastic.

In the 20th century, the paint industry has undergone dramatic changes. The advances in this century have outstripped all the advances made over the centuries before.

paint finishes The lustre, or sheen, that paints comes in.

High gloss paint has a high shine. Available for both inside and outside the house, it is commonly referred to as "enamel," implying particular durability, which is not correct. High gloss paint is normally used on interior and exterior trim and in kitchens and baths.

Semi-gloss paint has a sheen slightly lower than high gloss.

Satin-flat paint has a slight sheen comparable to the sheen of satin or an eggshell (another way it is described). Satin-flat paint is available for both interior and exterior use.

Flat paint has no sheen or gloss. Also known as matte flat paint, it is available for both interior and exterior use.

pallet Small section of wood inserted between BRICKS around a door to provide a nailing surface.

pane A single section of WINDOW glass.

panel 1. Non-loadbearing wall in frame construction built between columns or piers. **2.** In masonry, a panel wall is made of block and resembles panels in that it comprises the entire wall from floor to ceiling. It also resembles the CURTAIN WALL in that it is tied to steel framework, except that while the curtain wall rises from FOUNDATION to ROOF, the panel wall only goes from floor to ceiling high with each WYTHE of block. It is tied to the metal frames by flexible angles so the panel walls on one floor are independent of any above or below. This way, if something happened to a floor above or below in a three-story building, the middle floor would not necessarily be affected.

panel box Another name for fuse or circuit breaker box in a house.

panel clip Metal device for supporting panel edges to reduce deflection on roof construction.

paneled construction Building components fabricated in sections elsewhere, carted to a site, and assembled there.

paneling Sheet or board material used to finish walls.

Paneling comes in sheets and planks. Standard panel size is 4′ × 8′; plank size can range in thickness, widths, and lengths. There are a number of different sheet types.

Plywood sheets have a face of fine wood and, sometimes, exotic wood that can run the price up greatly without improving quality. Such woods may have a HARDWOOD core, which can make it difficult to cut, or a SOFTWOOD core, in which case it should not be. The material is either ⅛″–¼″, and the thicker material, the better. Another kind of plywood paneling has a vinyl face imprinted with a wood grain.

Hardboard is a 4′ × 8′ paneling with designs imprinted on the surface. Hardboard comes in a wide variety of colors and styles, many of which imitate wood.

Wallpaper paneling, one of the newer types of paneling, is sheet material with its face and edges covered with wallcovering. Installed, wallpaper paneling looks like wallpaper.

Planks are pure wood that range from 3″–12″ wide, come in lengths of 8′, 10′, 12′, and 14′, and in thicknesses from ½″–1″. Edges are TONGUE-AND-GROOVE.

Planks are either clear or knotty, the clearer material being the higher priced. Size also adds to cost. Knotty grades for paneling run from 1 Common to 3 Common with 1 the clearest. Clear grades go from superior pine (the clearest) to D Select.

As plaster came, wood went

In early America, wood planks were the most popular wall covering. After plaster appeared, wood was used less frequently to cover entire walls and more as accents, running perhaps from floor to CHAIR RAIL, or covering one wall.

Raised panels were part of this. Raised with single slabs of wood perhaps 3′ × 4′, they would "float" free inside a grooved framework of RAILS and STILES instead of being made immobile with fasteners. This was done to accommodate seasonal weather changes.

During humid weather, the panels would expand, and contract during dry weather. If the panels could not move inside the framework, something would crack and give. This is why later, when such panels were painted and, in effect, glued immobile, the change in humidity stresses often led to cracking.

parapet wall The part of a wall above the roof line. In modern construction, parapet walls often are part of MASONRY construction.

parging Applying a rough coat of plaster or masonry. Parging, formerly known as pargeting, from the French words *par*, meaning "throughly," and *jeter*, "to throw," is used to protect a surface from water penetrating it.

If there are two WYTHES (walls), the normal procedure is to apply mortar to the facing surface in what is to be the interior wythe, and then to erect the outer wall. If there is only one wall, the interior side of the wall will be parged. Another common method is to build a brick wall, parge the entire wall, and lay the BRICK against this.

Do not stop in the middle of parging a wall, even with the intention of returning to finish; a continuous, solidly adhering coat of mortar is required. If interruptions are unavoidable, the best solution is to feather the mortar where you stop. When you return, overlap it with a fresh coat and continue to finish the job.

Block basement walls should receive a double parging coat. The first coat should be ¼″ to ¾″ thick and troweled either vertically or horizontally.

The coat should go all the way to the base of the foundation, with a COVE joint drawn. This will keep water from standing at the base of the wall.

Once the first coat has dried, it is roughened up and wet down, and a second coat is applied. The second coat is applied in the opposite direction of the first; if the first coat was horizontal, the second should be vertical. This should be damp-cured for two days and then, for double assurance, an asphalt base coat should be applied over it.

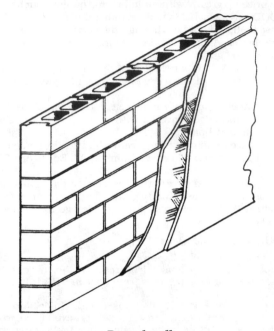

Parged wall

parquet flooring Hardwood flooring that comes in squares ranging from 9″ to 36″. Like STRIP FLOORING, parquet has tongue-and-groove edges. It comes finished and unfinished. *See also* PLANK FLOORING.

particleboard A heavy construction panel with bonded wood particles arranged in layers in random fashion and bonded with a phenolic resin. Some particleboard is structurally rated.

As materials go, particleboard is not easy to work with. It is rough on saw blades, and therefore difficult to cut; it cannot be EDGENAILED because it will not accept the nails; and, already heavy, it swells when it comes into contact with water.

Still, when it is used properly, particleboard is an excellent material, costing half of what PLYWOOD does.

partition wall A wall that subdivides space within a building but does not have a supportive function. *See* WALL.

party wall A wall shared by two separate living areas.

patio door Door with sections that slide back and forth. Patio doors, also known as SLIDING DOORS, came into vogue at the end of World War II, when family activities moved off the porch to the back of the house.

Patio doors may have two, three, or four panels. A two-panel door has one sliding and one moveable panel; a three-panel door, two fixed and one moveable; and a four-panel door usually has two operable panels in the middle and a pair of fixed panels flanking them. Patio doors come "knocked down"; the door panels and frame are packaged separately.

Patio doors may also have swinging panels. These are usually two- or three-panel doors. Typically, the operable panel is hinged to an inactive panel with the latch on one side of the door; in a three-panel door, the moveable panel may be in the center.

Patio doors, like regular DOORS, are made from various materials. They can be made in PVC, usually reinforced with metal; pure aluminum; vinyl-clad wood; or wood. Wood is generally top-of-the line. *See also* FRENCH DOOR.

pavers MASONRY materials formed into units and installed as paving material.

peak The top of a house.

pegboard Brand name for HARDBOARD with small holes punched in it.

penetrating finish Any of a variety of finishes where the product seeps into the pores of the wood. *See* STAIN.

penny A measure of NAIL size signified by the letter "d."

percolation test Different soils can absorb liquid to different degrees. Sandy soil has tremendous absorptive abilities; some hard clays have very little. Because of this, it is essential that soil be tested for its permeability prior to the installation of a system for disposing of sewage (*see* ABSORPTION FIELD). Additionally, the area may be full of rocks, become swampy when wet, or have a high WATER TABLE, all of which make the spot unsuitable. It can be costly for the contractor to build a dispersal system on soil only to find out that a larger, more elaborate, and costlier system is required.

A percolation test is relatively simple, although it varies from one BUILDING CODE to the next. After digging a 12"-diameter hole as deep as the DRAIN TILE will go, the contractor fills the hole with water and notes the absorption rate. Different areas have different criteria, but, in general, if 1" of water is not absorbed per hour, the soil is considered unsuitable for a septic system or cesspool.

permits Various written approvals one must get from a municipality before building can begin.

When you apply for what is commonly known as a "work permit," you must supply a scale drawing to indicate exactly what will be done, as well as the techniques and materials to be used. The plans *must* indicate that you are following BUILDING CODES. Other requirements differ from community to community. *See also* BUILDING CODE.

pick and dip Bricklaying method where the mason picks up a BRICK in one hand and uses the other to scoop up sufficient mortar to lay the brick.

picture rail Molding that runs around a wall close to the ceiling, designed for hanging pictures from. It is also used as a decorative strip above wallcovering, in which case the section above it is the frieze.

picture window A large, rectangular window with much glass. Picture windows come in a variety of styles; most common are BAY and BOW.

pier A column of masonry, usually rectangular or horizontal in cross section, that supports other structural members. *See illustration on following page.*

pigtail A method of tying electrical wire so it does not come loose. The name comes from the appearance of the tie, which is coiled like the tail of a pig.

pilaster A projection from a wall forming a column to support the end of a beam framing into the wall. *See illustration on following page.*

pillar Upright column of stone or brick used to support superstructures, but which stands alone for its slender good looks.

pilot hole A hole drilled slightly smaller than the screw that will be used in the hole. Pilot holes are used to prevent the wood from splitting.

pilot light Flame in an appliance fueled by gas.

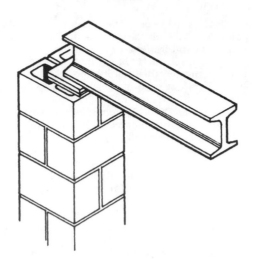

Pier

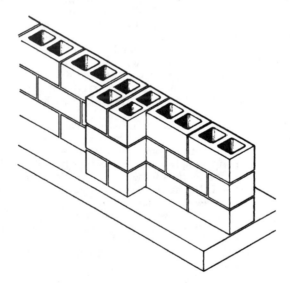

Pilaster in a block wall

Source of ignition

A number of building products are flammable. To avoid a fire or igniting gases, pilot lights must always be temporarily extinguished to avoid problems. *See also* FLASH POINT.

pitch Also called "incline," pitch measures the SLOPE of a roof. Pitch is expressed as the ratio of the rise of the slope over a corresponding horizontal distance. Measurements are taken in inches and feet. For example, 4 in 12 would mean that the roof rises 4″ for every horizontal foot.

When steepness was a virtue

A steeper roof is stronger at supporting snow loads and at shedding snow and water. Steep roofs can also use materials such as CEDAR, SLATE, and tile that could leak in lower pitch roofs. ASPHALT and fiberglass shingles and membrane roofs allow for much lower slopes.

Architecture in this century has been marked by less-steep roofs not only because of the materials available, but because of prefab roof TRUSSES.

pith The small, soft core at the original center of a tree around which wood formation takes place.

plancier Another name for an EAVE.

plank Commonly applies to a long, thick length of wood at least 1½″ thick and 6″ wide. In many cases, planks are much larger.

plank flooring Wide boards with TONGUE-AND-GROOVE edges that serve as flooring. Planks may be 3″ or wider; 6″ and 7″ inches are common widths, but planks can be wider. The planks are screwed to the floor with plugs inserted in the holes to produce a wood-peg effect.

Plank flooring comes finished and unfinished. As with STRIP FLOORING, factory processes result in boards that are fractionally different in thickness. This can make adjacent panels look different, but can be avoided by making the edges slightly grooved so that discrepancies are not noticeable. If needed, unfinished flooring can be sanded down to eliminate height discrepancies. All unfinished flooring must, at any rate, be sanded.

plaster Material used as a finished wall material.

Plaster consists of GYPSUM combined with water and LIME, sand, portland cement, or other materials, depending on the job. This mixture, a wet, highly plastic, and workable mass dries to a hard, fireproof, vermin-proof, and glass-like smoothness that accepts finishes beautifully.

Plaster was once the primary finishing material in America, but it has been replaced by DRYWALL. The reasons were not quality but economy. Plaster walls are better than drywall ones, particularly regarding sound transmission; but it is easier to screw home a 4 × 8 sheet of drywall and tape the seams than it is to apply the two or three coats of material necessary for plastering.

A plaster job may have two or three coats. If the base is solid, being made of MASONRY or gypsum board LATH, only two coats need to be applied: a BROWN coat and the finish or PUTTY coat. If the lath is wood or metal, three coats are usually applied: a scratch coat, a brown coat, and a putty coat.

In a three-coat plaster job, the first coat is applied and allowed to dry somewhat. After this, it is scored with a comb-like tool known as a "scratcher." When the scratch coat has hardened, the brown coat is applied; and, in due time, so is the putty coat.

plaster board *See* DRYWALL.

plaster ears *See* EARS.

plaster grounds *See* GROUNDS.

plaster lath *See* PLASTER.

plaster of paris Powdered material made from pure GYPSUM. When plaster of paris recombines with water, it becomes plastic but loses its plasticity and hardens in about ten minutes. There are "retarders" available to extend the setting time considerably. Plaster of paris has long been a favorite of painters, who use it to repair cracks in PLASTER walls. Its name comes from the discovery of a store of gypsum in Paris, France in the late 19th century.

plastic laminate Thin, hard plastic sheet material used for facing countertops and cabinets. The most common laminate is the brand name Formica, known in the trade as 'Mica.

Plastic laminate comes in widths of 2'–5', in lengths up to 16', and in a wide variety of colors and textures. Most of the material has a color facing and a brownish core. Installed, this brown line shows at the joint where the top of the countertop meets the edge. A more expensive laminate has the color all the way through.

Plastic laminate comes in several thicknesses. The standard size is 0.050 and is designed for use on countertops, but there is thinner material— 0.042 gauge—for use on cabinet faces, where it will not take the punishment a countertop gets.

Roll laminate—a type of laminate that is thin, flexible, and easily stained—is not normally used by professionals.

Installation of plastic laminate is not difficult. It is applied to a base of MDF, PARTICLEBOARD, or PLYWOOD. First, a piece of laminate is cut big enough to cover the counter. Contact cement—so called because it bonds on contact—is applied to the back and the base of the laminate and left until it is dry to the touch. Dowels are placed on the base at intervals, and the laminate set on the counter. The dowels are removed one by one, and the laminate is pressed down into place. Any excess is trimmed with a router. The procedure is repeated to apply an edge strip. Kraft paper may also be used.

Today's plastic laminates come in a wide variety of finishes, and are textured and tooled to resemble such things as slate. Some plastic laminate looks old-fashioned. To give it a modern look, some installers use wood strips along the edges instead of laminate. *See also* COUNTERTOP.

> **Making plastic laminate**
>
> Plastic laminate is made with layers of kraft paper that have been saturated with phenolic resins. It has an exterior colored layer of paper of a higher quality than the inner papers and is coated with a melamine resin. The sheets are pressed and bonded together under high heat.
>
> The first brand of plastic laminate on the market was Formica, which was introduced in 1913 as an electrical insulator. In the 1930s, Formica became a popular countertop material. It remains the most popular material for countertops to this day.

plasticity The workability of a material, such as PLASTER.

plasticizer A substance that makes concrete more workable with less water. Plasticizer also increases the concrete's strength because of the reduced water content, and reduces labor costs because the concrete is more workable.

plat A map of a section of land showing its length, width, and any EASEMENTS on it.

plate The flat members in STICK-BUILT structures. *See also* PLATFORM FRAMING.

platform framing Also known as WESTERN FRAMING. A form of stick framing (*see* STICK BUILT), it is the most common method of house framing. In platform framing, the framing members are secured to a floor, or platform.

Like balloon framing, platform framing begins by bolting a SILL PLATE to a concrete foundation or bed of mortar. The sill is nominally TWO BY. The next step is to lay JOINTS across the sill and spike them to it. Once this has been done, a header (which is not found in balloon framing) needs to be laid on edge and spiked to the sill boards. The resulting structure is a box sill. The final step in forming the platform involves laying the SUBFLOORING, usually PLYWOOD sheets, on the joists and box sill and securing it there.

Once the platform has been constructed, you should next lay a group of two-by-fours known as shoes flat and nail them to the platform around the edge and against the headers. The shoes are the bottommost members of the wall and are not present in balloon framing. Next, make a framework of studs, toenailed 16" on centers to a top plate. At the corners are posts, nailed together double two-by-fours to form a recess to accommodate wall material.

When this wall assembly is complete, it is raised up and the ends are toenailed to the studs and braced in place. In turn, each of the walls are assembled on the deck, raised up and similarly toenailed in place.

If there is a second floor, the procedure is repeated: a platform is built and walls are erected on it.

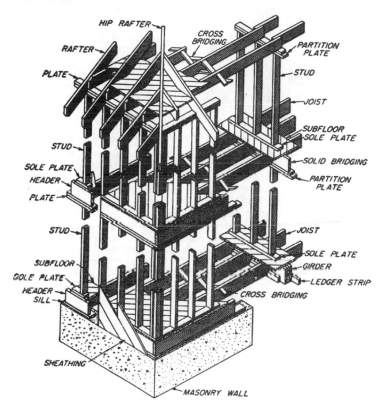

Platform, or western, framing

plinth 1. The lowest squarish portion of a COLUMN. **2.** The base of a wall of stones. **3.** Repair section in a PLYWOOD panel.

plug 1. Device on the end of an electrical cord that is inserted into an outlet. **2.** Small wood section inserted in a masonry wall to provide a nailing surface. **3.** A section of plywood inserted in a plywood panel to fix a defect. A plug may be circular or "dogbone" in shape. Synthetic fillers are also used as plugs. **4.** In electricity, the pronged device that is inserted in a receptacle and makes an electrical connection.

plug-in strip Electrical wiring encased in a shield with outlets for plugs. The strips are popular when installing standard OUTLETS is difficult. The strips are screwed onto the wall and a single connection made with an outlet.

plumb 1. Perfectly straight. **2.** Install plumbing. **3.** Perfectly perpendicular.

plumb and level Both perfectly straight and perfectly level.

plumb cut A straight up and down cut.

plumbing Collective term for pipes, FIXTURES, FITTINGS, and the like, that make up a plumbing system in a house.

How plumbing got its name

There is no great mystery to the etymology of the word plumbing—it comes from the Latin word, "plumbum," meaning lead. Originally, and for hundreds of years, pipe was made of lead, so it was natural to call the person who worked with lead a lead-er, or "plumber."

As time went by, the hazards of ingesting lead were discovered, and lead pumbing was banned. (Lead solder was still used for sealing joints in copper pipe, but this was banned in 1958.) Not all water is acidic enough to leach lead out of water, but it is always a good idea to consult with the local water company to find out what precautions, if any, are necessary in dealing with water supply pipes that might have lead joints.

Plumbing: an ancient art

Archaeological digs have come up with physical evidence that the principles of hydraulics and fundamental plumbing concepts were well known in ancient Crete. The plumbing achievements of the Cretans had been buried by natural disasters—earthquakes and possibly a volcanic eruption—and were only discovered in the twentieth century.

Archaeologists unearthed rudimentary but highly efficient water pipes used in the palace at Knossos on the island of Crete. These consisted of hollow pieces of terra cotta shaped like telescopes. The pieces were joined by sticking the narrow end into the wider one. Joints were sealed with a clay cement. The pipe could be made as long as wanted; additional strength could be ensured by lashing the pipes together with rope around knoblike projections. The water ran by force of gravity, and as it went, its action created a turbulent stream that washed away any sediment that might otherwise accumulate.

The same dig produced evidence of a drainage system complete with flush toilets with wooden seats, TRAPS, and VENT pipes to rid the system of harmful gases.

It was the Romans who developed plumbing on a sophisticated scale. They created a water supply and drain-waste-vent system used throughout their kingdom. In the fourth century, Rome boasted almost 900 public and private baths, 1300 public fountains and cisterns, and 150 toilets, all flushable.

Rome was a thirsty city, and to bring in the nearly 50 million gallons of water to meet the city's daily needs, a 359-mile-long network of aqueducts was created—some of which modern Rome still uses. The aqueducts operated above and below ground, the water borne along in CONDUIT until it was fed into smaller pipes (usually made of lead) that were buried underground.

The Roman plumbing system fell on hard times when Rome fell to barbarian hordes. The plumbing system gradually declined because there was little interest in the system and no one skilled to maintain it when it broke down.

From Roman times on, plumbing itself went into decline, borne along by beliefs that bathing was wrong, for one reason or another. Men and women went unwashed for literally hundreds of years. It was only in the 19th century that modern plumbing began to gain favor.

ply A term denoting one thickness of any material used for building up several layers. Examples of plies are roofing felt, VENEER in PLYWOOD, or layers in built-up material.

plywood Wood made of three or more layers of VENEER joined with glue, usually laid with the grain of adjoining plies at right angles to one another. This arrangement makes plywood strong and highly resistant to movement from expansion and contraction. This lack of cracking keeps paint and finishing problems to a minimum. Plywood has innumerable uses, from roof DECKING to SIDING and CABINETS.

A glue made it stick

Plywood has been around a long time. Its first use was probably in gluing wood shavings together to make furniture, but the real birth of plywood began around 1905, when Gustav Carlson of Portland, Oregon laminated thin sheets of wood together with animal glue and displayed his product at the World's Fair.

The first plywood panels, called "three-ply veneer work" smelled atrocious because of the animal glue used, but the product caught on. Panels were used for door panels, drawer bottoms, and even running boards for cars.

Sales increased steadily into the 1920s, but fell off during the Great Depression. This dropoff accelerated when car manufacturers started to make running boards from metal instead of plywood, which they had learned could delaminate when wet—animal glue was not waterproof. Still, plywood hung on, and, spurred by a series of daredevil stunts including a one-man trip down 600 miles of the Colorado River in a plywood boat, plywood sales rose.

In 1934, the real heroes of plywood emerged. Technicians developed phenol resins to glue the veneers together—resins that were waterproof. With this problem solved, plywood sales boomed. In World War II, plywood proved itself under fire. PT Boats were made of plywood, a sturdy light material that allowed the PTs to outmaneuver the heavier Japanese fighters, and GIs lived in plywood barracks and sat on plywood seats in bombers as they flew their combat missions. Plywood was here to stay. Today, although many other materials have been invented, it is still going strong.

pocket door A door that slides in and out of recess in the WALL.

pointed ashlar Tool marks made on the face of ASHLAR.

pointing Process of tooling MORTAR JOINTS in brick, block, or stone. Pointing is done to make the mortar joints look good, but it has a practical purpose as well. Working the mortar to a certain shape and compactness can give it greater or lower water resistance.

The term pointing comes from the procedure's being performed with a trowel point.

polish coat The final coat of JOINT COMPOUND when installing tape over drywall seams.

polyurethane Clear coating used to protect wood indoors.

Polyurethane, or "poly," is a commonly used protective coating for wood inside the house. It can be made with either an oil base or a waterbase, each with its own pros and cons.

An oil-based poly thins and cleans up with mineral spirits, a process that often takes overnight to dry. If the manufacturer recommends three coats for doing a floor, the job could take three days, allowing adequate drying time between coats. Oil-based polys yellow as they age, which some find objectionable.

On the positive side, oil-based products dry harder than water-based ones. This is particularly good for floors, which scuff readily with foot traffic.

Water-based polys thin and clean up with water. Unlike oil-based polys, they dry fast: it is normal to finish a floor job in one day because the poly can be applied every two hours or so. Additionally, water-based polys do not yellow; they stay clear for the life of the material. For vertical surfaces, unless you like the aged look, water-based polys would seem to be a good choice.

On the other hand, water-based polys cannot withstand floor traffic as well as oil-based polys. At least one manufacturer of water-based material (Carver Tripp) has recognized this problem and suggests that customers add a hardener supplied by the company to every gallon of poly they mix.

pond An incompletely drained roof surface. The name comes from the amount of water that can collect on such a surface.

porch Covered entrance to a house. Porches commonly have a roof distinct from the house.

porkchop Lingo for a LOOKOUT in CORNICE construction.

portico Structured like a roof with columns and a separate roof. A portico is sometimes attached to the main building like a porch but other times is completely separate.

portland cement Cement that sets and hardens by HYDRATION, a chemical reaction between a substance and water.

In a real sense, portland cement is like an adhesive. When water is added to it, it turns into a kind of paste. When aggregates and sand are added, it bonds them all together into concrete. This binding process is known as hydration, and it begins as soon as the water is added. The particles of cement bond to one other and to anything they are in contact with (the aggregates). As the water evaporates, the material loses its workability and forms CONCRETE. This process takes about three hours.

The manufacture of portland cement begins with stone; a combination of limestone, cement rock, or oyster shells; and shale clay, sand, or iron ore. The rock is reduced to pieces about 5″ in diameter, then to ¾″, and kept in separate storage bins. The rock pieces are combined in certain proportions and pulverized into a powder using the "dry method," or they are mixed with water and blended into a slurry. This raw mix is kiln-burned to partial fusion at 2700 degrees into what is known as "cement clinker." The final step is to add GYPSUM and grind the materials together to form portland cement.

The invention of portland cement

Credit for inventing portland cement is given to Joseph Aspdin, the English mason who patented it in 1824. He named it portland cement because the gray color of the concrete it produced resembled the color of the natural limestone quarried from the Isle of Portland, a peninsula in the English Channel. Calcareous cements, the family of cements to which portland cement belongs, had been used to make concrete for many centuries. Aspdin was simply the first person to patent it.

Portland cement did not have a large impact on building in America until the 1800s, as it was imported from Europe. The first recorded shipment was not until 1868.

post Vertical timber used in framing a building.

post-and-beam roof Roof consisting of thick planks spanning beams that are supported on posts. This construction has no attic or air space between the ceiling and the roof.

post-and-girt framing Way to frame a house, also known as TIMBER FRAMING.

post-formed countertop COUNTERTOP covered with a continuous, seamless sheet of PLASTIC LAMINATE.

potable Water free of impurities that could produce harmful physiological effects. Potable water conforms to the drinking water standards of the U.S. Public Health Service or the regulations of the public health authority having jurisdiction in a particular area.

pot life The amount of time an adhesive or other material remains usable.

poured-in-place insulation This insulation is poured into open cavities in the house, chiefly an open attic floor. It is also known as "loose" insulation.

There are two kinds: cellulose and fiberglass. They work essentially the same way. They come in bags that are opened and poured into place, and raked or leveled with a board.

See also BATTS, BLANKETS, BLOWN-IN INSULATION, *and* BOARD INSULATION.

precast concrete Concrete sections that are made, or cast, at one location and transported to another for use.

prehung door Door that comes already secured to its framework. It also comes with hole drilled for the lock.

preservative Chemical in coatings used to provide protection against insects and fungi. *See* STAIN.

pressure balance valve VALVE that keeps water pressure constant. This valve is useful for keeping hot and cold water at the temperatures they are set for. If the temperature changes even slightly, a diaphragm inside the FAUCET reacts instantly to reduce or increase flow as necessary to keep the flow constant.

pressure regulator Device used in a water service line entrance to reduce water pressure to something desirable for household use.

pressure tank Device used to pump water from a well.

pressure-treated lumber Wood injected with chemicals to make it resistant to insect infestation and rot.

The preservative is usually chromated copper arsenate (CCA), which gives the wood a slightly green cast. Each component has a specific function: copper acts as a fungicide, arsenic as an insecticide; and chromium, the agent that binds everything to the wood.

A treatment plant supplies the preservative as a liquid concentrate that is diluted with water. Once at the appropriate level, the preservative is injected into the wood in large steel treating cylinders at about 150 pounds per square inch. In order to accept the preservative solution, green lumber is usually dried to a moisture content of 25 percent or less by KILN-DRYING or by seasoning in the open air.

Once it is inside the cellular structure of the wood, the CCA treatment undergoes a complex series of chemical reactions with the major wood components—cellulose, hemicellulose, and lignin. These reactions bond the CCA to the wood fibers, rendering these chemicals insoluble and resistant to water leaching.

After the wood has absorbed all the treating solution it can and is completely saturated with liquid, the pressure drops and a short vacuum removes any excess liquid. The wood is then air-dried or kiln-dried before it is shipped to the lumberyard. The highest grades of wood are those that are kiln-dried to bring the liquid level down to nineteen percent. It is more typical for the wood to air-dry.

In some cases, treated wood will reach the lumber yards still wet. Although the preservative is fixed in the wood, excess residual moisture can have a negative impact on finish coatings; therefore, the wood should be surface-dry before it is coated.

Many types of softwoods can be pressure-treated with CCA. The most frequently treated species is southern yellow pine. In the west, hemlock, hem-fir, ponderosa pine, jack pine, and red pine are also subject to CCA treatment. Some species, such as DOUG FIR, have difficulty accepting treatments. It is sometimes necessary to incise the wood before treatment in order to facilitate the preservatives' penetration.

Treated lumber often bears a mark that indicates the level of CCA treatment. The CCA level is a retention number that represents the pounds per cubic foot of preservative in the wood. The level selected depends on the final use of the wood. For above ground applications, the specified retention level is 0.25 pounds per cubic foot (pcf) of preservative in the wood. For wood in contact with the ground the recommended level is 0.40 pcf; and for treated plywood, it is 0.60 pcf.

One misconception has it that pressure-treated lumber can resist just about anything. In fact, it resists only fungi and insects, not the ravages of weather. In fact, the weather takes its toll on pressure-treated wood as fast as it will on any other wood, and it should be treated for weatherfastness immediately. (*See* STAIN.)

Pressure-treated wood can be worked like any other wood, but it is dangerous to cut without wearing a mask or other protective device. Sawdust from pressure-treated wood is essentially arsenic-impregnated dust. There is a controversy about how safe it really is.

Another drawback of pressure-treated lumber is that it bows and bends more than non-treated lumber when it dries. This is probably because it is saturated with chemicals, and many lumberyards tend to store pressure treated in the open, where it takes on even more moisture.

Pressure-treated lumber is available in standard lumber sizes and as plywood, which is useful in making foundations (*see* FOUNDATION). It has found wide acceptance in the U.S.

How safe is pressure-treated lumber?

The answer is probably that no one knows for sure. It certainly is toxic to work with when sawn. It is still a question whether the poison leaches out over time.

primer 1. In painting, a first coat designed to seal the surface and provide good adhesion for finish paint.

There is a wide variety of primers available for both interior and exterior wood. In general, they are broken down into oil and latex types and interior and exterior, but there are also shellac-based kinds for use on water stains, tobacco stains, and the like. Sometimes, primers and finish paints are thought of in the same way, but they are different. **2.** Any of a number of first coatings for various materials. In painting, there are many primers. In asphalt roofing, one may apply a think coat of BITUMEN for adhesion.

Why primer?

Primers are better than finish paints in many situations because of three factors. First, they are more fluid than finish paints and penetrate the SUBSTRATE better, filling in cracks and crevices to some degree. Primers also provide a better surface for the finish paint to bond to. Primers also provide a nonporous base that topcoats cannot penetrate, something that could affect the look of the final coat.

profile An outline drawing of a section of a building component.

property line The boundaries of a property. These must be known so that a new building, addition, or the like does not encroach on someone else's rights. The author had a personal experience where knowledge of his property line proved useful. A contractor had told him exactly where he planned to install a new cesspool. The problem was that it was within 10′ of an adjacent road—right in the middle of town property. *See also* EASEMENT RIGHTS.

puddle In MASONRY, the agitation of freshly poured CONCRETE with a stick or the like to help eliminate voids or air bubbles. Also called "rodding."

pugging Coarse MORTAR laid between joists to reduce sound transmission.

purlin A horizontal timber supporting the common RAFTERS in roofs. *See illustration on following page.*

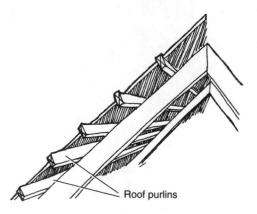

Roof purlins

Purlins

putty General term for a variety of soft, plastic materials used to seal or fill building products. Following is a lineup:
- **Electrician's putty** (*See also* MONKEY S--T.) Used to fill gaps where CONDUIT passes through openings.
- **Painter's putty.** Used to fill small holes.
- **Plumber's putty.** A bead of this is laid around the opening where a sink is to go. When the sink is set in place, the putty acts as a seal. Plumber's putty is also used as a sealant for installing FAUCETS.
- **Wood putty.** Material used to fill gaps in woodwork or cover COUNTERSUNK nails and screws. Wood putty comes plain and in various wood colors.
- **Window putty.** Formerly used to seal glass in a frame, but now displaced by GLAZING COMPOUND.

putty coat The final coat in a PLASTER job.

Applying the putty coat in a plaster job is *not* for a beginner. The material used, three parts LIME to one part GYPSUM, is a soft, plastic mass that must be applied to leave a hard, seamless, unmarred finish. This takes great experience—not only for knowing how to apply it, but when. The BROWN coat below must contain a certain amount of moisture, or it will suck moisture from the putty coat and weaken it. The putty coat is also known as the cream coat, white coat, hard coat, and smooth coat.

quarry tile Tile made from shale, clay, or earth, resulting in an unglazed tile with color throughout.

Quarry tiles are used for flooring as well as for walls. They are normally used indoors but may be used outdoors if the weather stays above freezing.

The term *quarry tile* covers a wide range of quality. Some tiles made with earth are soft and irregular and break easily. Others are so soft that they need a seal coat before they can be used.

quarter round MOLDING that is one quarter of a circle in profile.

quatrefoil A four-leafed decorative accent on a building.

queen post Vertical supporting member in post-and-beam roof construction.

quoin A projecting right angle masonry corner.

rabbet A rectangular, longitudinal groove on the corner edge of a board or plank.

rabbet joint The joint made when the end of a board fits into the groove made in a plank.

racking Installing roof shingles vertically rather than diagonally. Racking takes less time to do, but it is not recommended because it can result in random color variations and shingles that are nailed improperly.

rafter One of a series of structural members on a roof designed to support roof loads. The rafters of a flat roof are sometimes called roof joists. There are several varieties of rafter:
- **Fly hip.** A rafter that forms the intersection of an external roof angle.
- **Tail.** The part of a rafter that overhangs a wall.
- **Valley.** A rafter that forms the intersection of an internal roof angle. A valley rafter is normally made of double two-inch-thick members.

See illustration on following page.

rag work In masonry, building with small, thin stones in rubble work.

raggle 1. A groove in a joint. **2.** Masonry grooved to receive roofing or flashing.

rail 1. Cross member of a panel door or WINDOW sash. **2.** The upper or lower member of a balustrade or staircase extending from one vertical support, such as a post, to another.

rake Extension of a GABLE roof beyond the end of a house.

rake molding MOLDING secured to the rake edge of a building.

rake out Removing excess mortar from brick prior to POINTING it.

raking bond In masonry, laying brick courses in a zigzag fashion, as seen in the end walls of Colonial homes. *Also known as* "RAKE."

random tab shingles ASPHALT or fiberglass ROOFING SHINGLES with extra tabs randomly applied so the roofing resembles wood shingles.

random work In MASONRY, work done in an irregular fashion, such as building a wall with irregular stones.

181

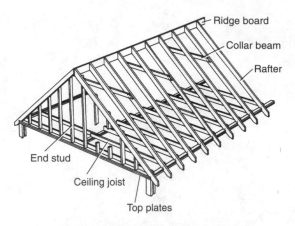

Rafter framing, gable-styled house

ranger Horizontal framing member, also known as a whaler.

ready mix Concrete that is delivered to a site ready to be poured. It is transported to the site by trucks that mix it en route. Concrete trucks are also known as "agitation" trucks. They carry nine cubic YARDS of concrete, hence the phrase that someone went "for the whole nine yards."

rebar Metal bar sunk in wet concrete to reinforce it.

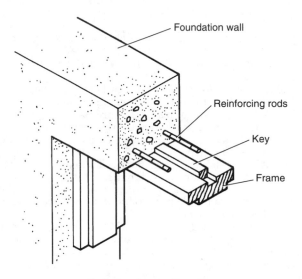

Reinforcing rods-rebar

receptacle Slotted electrical device that other devices are plugged into to draw power. Commonly called OUTLETS or sockets, receptacles are always "on"; once a device is plugged in, a CIRCUIT is completed and power flows.

    Receptacles are available in single or duplex style (two outlets in one device). They have terminal screws. The line, or HOT side screw is copper- or brass-colored; the neutral side is silver. Receptacles may also have a green grounding screw in one corner of the frame that attaches to the bare GROUND wire on the CABLE.

Most receptacles today also have holes in the back into which bare ends of wires are pushed. The holes are identified according to electrical function, which makes for a faster connection than the screw type does. These holes are also handy for wiring devices where there is not enough room on the screw terminals for the wires. The clamp holes allow for more convenient routing or wires within the box.

Like other electrical devices, receptacles are rated in terms of VOLTAGE and AMPERAGE capacity. Any device plugged in should not exceed this capacity. Standard receptacles are for fifteen amp currents, but "specs"-grade receptacles are available that carry twenty amps. These are better made than ordinary receptacles and are a good choice in a kitchen, bath, or wherever the receptacle gets a lot of wear.

Most receptacles have three slots—two flat slots for the plug prongs and a D-shaped one for the ground prong on the plug. All receptacles are—or should be—grounded. One type is automatically grounded when it is attached to the BOX. On other types, a wire must be fastened from the screws to the box. In cases with no easy way to hook grounding wires, a U-shaped "grounding clip" is attached to the box; or a green 8×32 grounding screw is mounted in one of the tappings in the back of every metal box for the wires to link into.

A "duplex" receptacle has openings to accommodate two plugs at a time. A square built receptacle is the same as a standard receptacle except that it's overall shape is square. This receptacle is available in fifteen- and twenty-amp sizes. Also known as decor receptacles, these are gaining favor because of their better appearance.

Receptacles may or may not have EARS to hold it flush with the wall if necessary.

Single electrical receptacle

See also illustration on following page.

recessed lighting fixture Lighting fixture recessed into a ceiling so its lip is flush with the ceiling.

register A device for controlling the flow of warmed or cooled air through an opening.

reinforced concrete Concrete with steel rods or metal fabric placed in it for strengthening. *See illustration on the following page.*

reroofing Applying new roofing material over existing material.

resilient flooring Flooring that "gives" when stepped on.

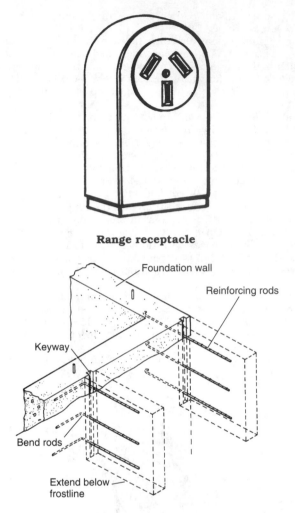

Range receptacle

Reinforced concrete

Resilient flooring comes in two basic forms: sheet goods and tile. The former is available in 12′-wide rolls of virtually any length. Tiles come in 12″ squares. The quality of these floorings, like many other building materials, varies considerably.

Vinyl sheet goods come in two forms: "inlaid" and "rotovinyl." Inlaid sheets come in a wide variety of patterns and colors. Because the pattern and color go all the way through the material, it stands up well to wear. Vinyl sheets are available in a "no-wax" finish. Inlaid sheets are glued in place.

Rotovinyl sheet goods are not nearly as good as inlaid ones. The pattern and color do not go all the way through and can wear through. The thicker this "wear layer" comes, the better. Sometimes rotovinyl sheets have a urethane coating and/or a cushion backing. Rotovinyl can be installed loose or by placing glue around each sheet's perimeter.

Thin vinyl tiles wear out quickly. With this in mind, thicker tiles are better. They come in thicknesses of one sixteenth, three thirty-seconds, and one eighth of an inch.

The key to installing vinyl flooring is the preparation. The base, or substrate, must be smooth, dry and sound. (*See also* UNDERLAYMENT.) Tile may be either the self-stick type or "dryback," installed without adhesive.

restrictive covenants Restrictions on the use of real property. Such covenants are created by deed and may "run with the land," meaning that no matter who owns the land the covenants must be observed; or they may exist only between the original buyer and seller. Just which is determined by state law, intent of the parties, and the language of the covenant. Restrictive covenants are also ENCUMBRANCES.

retaining wall A wood, BLOCK, or CONCRETE wall designed to alter the topography or provide improved storm water management; i.e., hold back the earth and water after rain so that they do not impinge on living area.

Retaining walls must be constructed to resist the pressure of earth and water, which, in effect, are trying to push it over all the time.

It would be foolhardy to try to design a wall that would resist this pressure totally without giving it some relief. One approach is to give the water a way to escape through the back of the wall. With a wood wall, this can be done with a gravel bed. MASONRY wall can utilize WEEP HOLES. If the wall is made of masonry, it must be protected from being saturated with water. This can be done by PARGING it with two coats of cement plaster or MORTAR. If the masonry becomes saturated, the water may solubilize salts in the masonry, leaching them to the surface and making the wall unsightly. *See* EFFLORESCENCE.

The classic wood retaining wall is constructed of PRESSURE-TREATED LUMBER. How it is built depends on the area. In areas with well-drained soil, the bottom most members may be placed directly on the soil. If the soil is not well drained, 12"–24" of gravel behind the wall and a 6"-deep gravel FOOTING are best.

The members are spiked in place with staggered JOINTS. Each course is nailed to the course below it with galvanized spikes one-and-a-half times as long as the timbers are thick. To prevent the wall from tipping forward, a combination of "tieback" and DEADMAN TIMBERS should be used. Every second course should include tieback members placed perpendicular to the face of the wall and spiked to the lower course. Each should be as long as its distance above the base of the wall, and each end of each tieback should be nailed to a dead man timber 2" buried in the soil parallel to other members. Tieback and dead men should be installed every 4'–6' along the wall.

One common kind of block wall is the "cantilever" type. For these walls, a footing is installed that is wider than the block. The footer is heavily reinforced and buried deep in the earth, and the reinforcement bars have the block slipped down over them. This REBAR can vary, and include block specially made for retaining walls, but how strong it will be depends on the architect or builder who designs it.

The main advantage to building such a wall is its solid mass. This is done with the rebar, which ties the block to the footing, which is buried in the earth. To push the wall over, the earth and water are not just pushing against the wall but against everything—the footing, the wall, and the ground that buries the footing—a task considerably more difficult than pushing down the wall itself. *See illustrations on following two pages.*

return corner block Corner block in the shape of the letter L, designed for constructing corners in six-, ten-, and twelve-inch block walls.

reveal The slight space between a door frame and the wall it abuts.

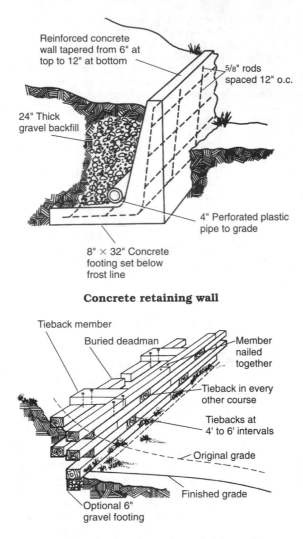

Concrete retaining wall

Pressure-treated wood retaining wall

ribband In BALLOON FRAME construction, horizontal members which support the second floor joists. *Also known as* "RIBBONS."

ridge The horizontal line at the junction of the top of two sloping roof surfaces.

ridge beam A heavy timber placed along the top of the roof, to which the RAFTERS are fastened. It is also known as a ridge pole and ridge board, although there are some differences. A ridge pole is a narrow board used in rough construction, and a ridge board uses an ordinary plank. Years ago, builders did not use ridge beams at all; they merely pinned the ends of the heavy rafters together. *See illustration on following page.*

ridge capping The line or course of material that covers the RIDGE BEAM.

ridge pole *See* RIDGE BEAM.

rigid frame construction Construction in which structural members function like an ARCH; composed of STUDS and RAFTERS fastened with PLYWOOD

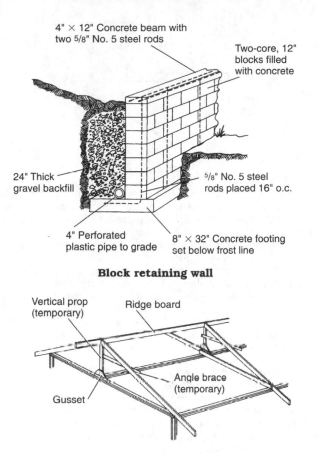

4" × 12" Concrete beam with two 5/8" No. 5 steel rods

Two-core, 12" blocks filled with concrete

24" Thick gravel backfill

5/8" No. 5 steel rods placed 16" o.c.

4" Perforated plastic pipe to grade

8" × 32" Concrete footing set below frost line

Block retaining wall

Vertical prop (temporary)

Ridge board

Angle brace (temporary)

Gusset

Ridge beam

GUSSETS. Rigid frame construction eliminates the need for ceiling or tie members.

ripoff Tearing off roofing materials down to the DECK, or wood base. Another name for TEAROFF.

rise 1. The distance from the attic floor to the ridge or roof peak. **2.** In stairs, the height of a step or flight.

rise and run Phrase used to describe the relative steepness of a ROOF; rise is height and run is horizontal distance.

riser 1. Vertical pipe in a plumbing system. *See* DWV SYSTEM *and* WATER SUPPLY SYSTEM. **2.** Vertical piece on stairs. *See illustration on following page.*

rock Lingo for Sheetrock, a common brand name for DRYWALL. *See also* ROCKERS.

rockers Term for people who install Sheetrock.

Romex One trade name for non-metallic CABLE. Romex is to wire, namewise, as Vaseline is to petroleum jelly. *See* WIRING.

roof All the material—framing members, DECKING, and covering—that is installed on the top of the house. Other than aesthetics, a roof has two main functions: to keep the interior dry and to insulate—keep the warm or air conditioned air inside.

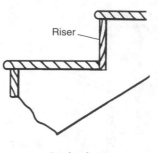

Stair riser

Roofs may be PITCHED (SLOPED) or FLAT, depending to a large degree on the area of the country. In areas with heavy rain or snow, pitched roofs are a must if the roofs are made of materials such as asphalt shingles, ceramic tile, slate, or wood is used. Roofs made of such materials are not impervious to water, and if they are not slanted, they will not shed the water, which can then work its way into the house.

Flat roofs, also called shed roofs, are practical only where precipitation is minimal. Even so, they should be covered with some waterproof roofing and be pitched to some degree to shed water. Water left to evaporate on a flat surface can cause problems, no matter what the covering.

There is a good variety of pitched roofs, each with its own features. A pitched roof might be gambrel, hip, gable, or mansard.

A gambrel roof, viewed from the front, has a hut shape. The roof looks like the walls have been extended and bent at one point. This roof is common for farm buildings; its design allows for maximum storage space, good for storing feed for animals.

A gambrel roof consists of lower rafters that angle off the PLATE and join at their ends to the upper rafters. The sets of rafters are nailed and tied together with horizontal boards called PURLINS. Besides the tying-together function, purloins help the roof to resist wind and weight on the roof and to stiffen the overall structure.

The hip roof is four-sided, the sides equally sloped. Hip roofs usually have a RIDGE BOARD running down the spine of the roof and "common rafters" running between the ridge and the wall plates. On the short sides of the hip roof are JACK rafters, short rafters that lay between the end common rafters and the top plate. The common rafters are made of boards that are wider than the jack rafters. The ends of the jacks have a full bearing surface because their ends are cut at an angle, making them deeper than they ordinarily would be.

Homes less than 20′ wide with hip roofs often do not require COLLAR BEAMS to stiffen construction, but they do require JOISTS at the ceiling to strengthen the structure.

The gable is the simplest roof of all. It consists of a RIDGE board that leans against and is tied to the common rafters. The common rafters have BIRD'S MOUTH cuts at the end opposite the ridge board, where they meet the top plate. collar beams tie the common rafters together, as do ceiling joists that extend across and connect the rafters at the bottom ends. The gable roof gets its name from its triangular ends, which simply consist of end STUDS that fit between the top plates and ridge board.

The mansard roof design is a blend of hip and gambrel roofs. Like a hip roof, it has four sloping surfaces, but it is bent halfway up like a gambrel. The main asset of the mansard roof is the storage space; like the gambrel, it is essentially a series of sloped walls.

Mansard roof design was popular in big Victorian houses. Like many other ideas, it came from Europe. Specifically, it is an adaptation of the French Empire style of architecture, which was popular in the 1850s.

Flat roofs may be slightly angled, dead flat, or slightly raised in the center to ensure water runoff. A PARAPET may or may not run around the perimeter.

Flat roofs are framed like floors, complete with joists the ends of which rest on TOP PLATES. If there are openings in the roof, they are framed as headers are framed. The framing is designed to support weight: the weight of the roofing material and the weight of snow loads. If the framing is not strong enough, the roof will collapse, as one occasionally hears about.

Slightly pitched roof

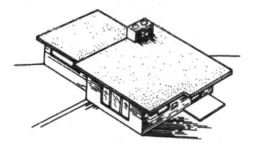

Flat roof

roofing A variety of materials made for flat and pitched roofs, including ASPHALT ROOFING SHINGLES, clay roofing tile, and concrete tile.

Clay tile is made from materials that are fired at high temperatures. Tiles come in various shapes and are made to interlock as well as lie flat. The tiles usually come in earth tones; but bright blues, reds, and other colors are available for an additional cost. Clay tiles also come glazed. These tiles are heavy.

Clay tiles are secured to the roof with any of a variety of corrosion-resistant nails. Installation proceeds as with any other roofing: tiles are installed, starting from the eaves and working up toward the ridge. When installing them, the craftsman will be particularly aware of flashings.

Concrete tiles look like clay tile and are available in the same shapes, but come unglazed. Like clay tile, they have holes punched in them so they can be nailed into place.

Elastomeric membrane roof, also known as "rubber" roofing, is a pliable roofing material. Commonly incompatible with asphalt roofing, if it is treated with a CONTACT CEMENT-type adhesive, elastometric membrane roof can be applied in large sheets. Mechanical fasteners are also used to finish securing it. The most popular elastomeric membrane roof is ethylene propylene dine monomer (EPDM), which comes in sheets up to 20′ wide.

Roll roofing is composed of fiber and saturated with asphalt. Applied in thirty-six-inch wide rolls of 108 square feet, it weighs 45 to 90 pounds per roll. The heavier the material, the better. Roll roofing is popular for flat roofs or roofs with a SLOPE of less than "two in twelve." On roofs where standard shingles—such as ASPHALT—are used, water sometimes freezes before it runs off. As it melts, it can back up under the shingles and damage the roof, something prevented by roll roofing. Roll roofing can also be used as FLASHING.

Slate is the most expensive roofing material available. It comes in a variety of grades and colors—gray, maroon, black—depending on the quarry the slate comes from. It mostly comes from states in the northeastern US. (Slate from Maine and Vermont is considered the finest.) Higher grades of slate have a grain that runs along the length of the slate; they are stronger than lower grades of slate; and are the slate with color. Coarser material may streak colors, a problem that is less likely with higher quality slate.

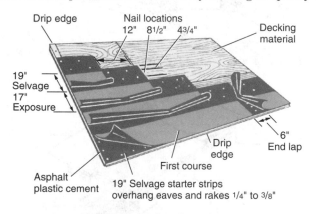

Details of roll roofing installation

roof boards Boards secured to RAFTERS as a base for roofing materials. These days, plywood and other sheet materials are mostly used for DECKING, but boards are still used as the base for wood roofs, which must be allowed to breathe.

rough in Installed plumbing pipe before fixtures are connected to it. *See illustration on following page.*

rough lumber Lumber that has been cut at the mill, but not DRESSED.

rough opening In general, openings made in a structure before the other framing members or trim is installed; an unfinished DOOR or WINDOW.

rout A technique involving making a groove in something, or routing out unwanted material.

routing Cutting away excess material. When a PLASTIC LAMINATE is installed, a common way to trim material that overlaps a counter is with a

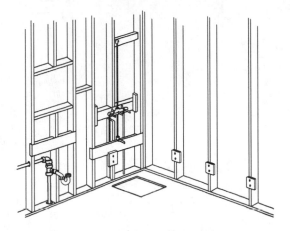

Rough–in plumbing

router. A carbide bit is run along the edge of the counter, neatly trimming the excess away. Routing is also used to trim excess PANELING that over-laps windows and doors.

rubber roof *See* ROOFING.

rubble Random stone used before manufactured MASONRY materials were available. The stones were assembled into a wall one by one, a laborious process with some quite beautiful results.

R-value The heat resistance value of a material. This is a key concept in INSULATION. Different materials have different R-values; the higher the value, the better the material is for insulating purposes.

saddle 1. The beveled board across a doorway. 2. Two sloping surfaces meeting in a horizontal ridge, used between the back side of a chimney or other vertical surface, and a sloping roof. Saddles are also called CRICKETS.

sag In painting, the running of PAINT once it is applied. A number of things can cause sag, including paint that is too thin, overapplied, or not brushed out. Such paint can eventually peel.

sandstone A sedimentary rock. Like limestone, sandstone can easily split along the sediment lines, making it an easily usable material for veneer and walkways.

sapwood The outer zone of a tree next to the bark. In a living tree, sapwood contains some living cells (the heartwood contains none) as well as dead and dying cells. In most species, sapwood is lighter colored than the heartwood. In all species, it lacks resistance to decay.

sash *See* WINDOWS.

sash weights Cylindrical weights used to counterbalance the weight of a sash in a DOUBLE-HUNG WINDOW. SASH weights make windows easier to open and close. Cords woven to resist stretching are normally used to hold the weights.

S beam Same as I BEAM.

scab Short length of wood nailed to two other members to splice them together. *See illustration on following page.*

scaling Pulling off the tabs of ASPHALT ROOFING SHINGLES preparatory to installing a new roof. The tabs, which resemble fish scales, are taken off because they are bent or curled and doing this allows for a suitable base for new roofing without having to TEAR OFF all of the roof.

scarf joint Joining boards by BEVELING the ends to fit together and fastening them with bolts or the like. *See illustration on following page.*

sconce Decorative bracket that projects from a wall and holds candles.

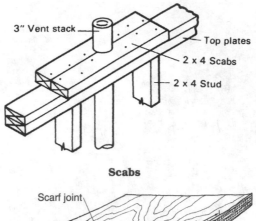

Scabs

Scarf joint

scoring 1. Making grooves on a material to improve its ability to adhere another material to it. **2.** Decorative grooves in the face of masonry.

scratch coat The first, or rough coat, in a PLASTER job.

screeding Leveling fresh concrete by drawing a board back and forth across it, using the FORM tops as a guide.

screen Woven material installed in doors and windows to allow the passage of air while keeping out insects. Screening can be aluminum, anodized charcoal, bronze, or fiberglass.

Aluminum screens come in fine (18 × 16) and heavy (18 × 14) wire. The former is for ordinary situations; the latter, for screens that might be subject to abuse, such as from children. Aluminum screening reflects light better than any other type of screening; it is relatively opaque from the outside.

Anodized charcoal is black aluminum screening, the most transparent of all. It is available in the same sizes as aluminum.

Bronze-colored metal screening is the only one that will last a lifetime in areas close to saltwater.

Fiberglass looks like any other screening; it comes in gray or black. Fiberglass is the least expensive screening there is, and can last for years, even in saltwater areas. Its only drawback is that it is easily cut; objects bounce off aluminum cut fiberglass.

screen molding MOLDING that goes around a screen frame and holds the edges of the screening in place.

screen wall BLOCK wall with decorative perforations which allow some privacy but also provide something of a barricade to wind and sun.

screw Threaded metal rod with a point and a head for turning. Screws generally come with a straight slot or "Phillips" head, a crisscross slot. Screws are designed to operate best in certain materials, although some do have multiple uses.

Screws come with a variety of different characteristics that must be considered when selecting a particular type for a job: finish, length, gauge, and head. Screws can be plain steel, blued, or dipped. The latter two—which include all galvanized, brass, and chrome-plated screws—are generally resistant to moisture.

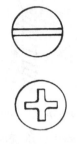

Screws: straight-slot and Phillips screw heads

Screws are classified by gauge, ranging from No. 5 to No. 14, although larger sizes are available. The gauge refers to the diameter under the head. If the screw tapers, the diameter is smaller at its tip than near its head. Screws of the same gauge are available in different lengths.

Contek screws are excellent for securing boards to concrete. Many carpenters consider then safer to use than MASONRY NAILS. Contek screws have double-pitched threads, threads that go at different diagonal directions, with a recessed Phillips or $5/16$ hex head. They are zinc chrome-plated. These screws must be installed with a masonry drill bit and electric drill.

Contek screw

Drywall screws are made of blued steel to keep them from rusting in the box. Threaded all the way, drywall screws range in length from $1\frac{1}{2}$" to 3". They have flared-out heads that resemble the end of a bugle, and are sometimes referred to as bugle-head screws.

Drywall screws are effective at preventing nail popping, the common malady that occurs when installing drywall. Nail popping can occur for a variety of reasons, including green wood shrinking and pushing the fastener out. Screws stick tenaciously, preventing this problem quite nicely. *See illustration on following page.*

It is common practice to drill PILOT HOLES when a lot of small screws need to be driven into wood. This is usually done with a screw gun, the very approach for installing drywall screws. On one type of screw gun, the tip is

Drywall screw

magnetized. The screw is placed on it, inserted into the pilot hole; and a squeeze of the trigger drives it home. Some contractors also use an automatic screw gun that reloads itself. When one screw is used, another one drops into place.

The machine screw is threaded along its entire length. The top may be flat or round, and the screws come in plain and CHROME-PLATED finishes, as well as brass. Threads on the screws may be coarse or fine and vary in length from ½"–4". Machine screws are used to assemble metal components.

Machine screw

Sheet metal screws are tapered along their entire length to a pointed end. They have loosely spiraled threads for securing metal components such as sheet metal. They may have either slotted pan heads or Phillips heads. They cut their own threads through metal, but drilling an undersized pilot hole first is recommended.

A wood screw has a tapered, partially threaded shank that comes to a point. It may have a flat, round, or oval head. (The roundhead is the easiest to grip and turn.) As the name suggests, wood screws are used for assembling wood components. They are usually available in brass and steel.

In some situations, screws are countersunk; their heads are driven beneath the surface, and the resulting depression is filled with wood putty. A countersink bit is used to accomplish this. This has a profile that allows it to drill a pilot hole for the screw and the depression for the head at the same time.

screw tek Sheet metal screw used to drill and fasten heavy sheet metal in one operation.

These screws are fully threaded with a tip shaped like a drill bit and a head that is either Phillips or hex. Screw teks come zinc-plated in various sizes.

Where the Phillips head screw comes from

The Phillips head screw is named for its inventor, Henry F. Phillips of Portland, Oregon. The screw won him a patent in 1934, and was offered for sale in 1935. As reported in the November 7, 1935 issue of *Iron Age* magazine: "American Screw Company announces a line of case-hardened sheet metal screws featuring a new Phillips recessed, self-centering head in place of the conventional screw-slot. The geometric patterns of the Phillips head provides that the screw shall hold to the taper point of the driver and may be brought into position with one hand."

Phillips' driving force was to create a better screw, one the screwdriver was not likely to slip out of. To this day, the Phillips head screw is considered superior to the straight slot, and it is used for a wide variety of metal and wood applications.

screwed off Lingo for holding some material in place with a few screws, then screwing all the screws in at once.

scutcheon Metal plate used around a door keyhole to protect the wood.

sealer General term for any liquid material designed to seal for a particular purpose.

A sanding sealer is a light coat of clear finishing material used to seal a floor before applying a finish coat. Sanding sealers are essentially a watered-down version of the finish coating.

A concrete sealer is a clear coating that might be designed just to reduce dust on concrete to provide protection against moisture. There are clear sealers for use outside to protect masonry and wood from water penetration, although they are not waterproofers, which would keep water out of a basement even when under pressure.

seasoned wood Wood that has been allowed to dry outdoors.

self-siphoning A condition in a trap that leads to a compromising of the water seal in a TRAP.

Air pressure is the key to keeping a water seal in place. If this is compromised, self-siphoning can occur. Because there is no air pressure at one end of a tube to balance it out, the water rushes down the pipe in a solid stream. Eventually, the water will stop running, but as a result of the siphoning, the water seal is lower than it should be, allowing gases and vermin back through the top of the seal.

Self-siphoning can occur in any U-shaped section of pipe with legs of unequal length. This is why S TRAPS, which have legs of greatly different size, have been banned in most American communities.

The condition can also occur in P traps, but this is much rarer.

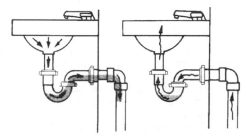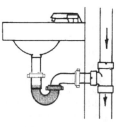

Self-siphoning

septic tank The part of a septic system that receives sewage and processes it for distribution into the ABSORPTION FIELD. The sewage entering the tank consists of both solid and liquid waste. Heavier particles sink to the bottom of the tank, becoming part of the sludge, while others disintegrate into lighter particles and rise to the top of the liquid.

The bacteria within the tank live in the absence of oxygen, where they break down the sludge into effluent, a white, toxic, bad-smelling liquid. When new sewage enters a full tank, an equal amount of effluent discharges to the drain tiles, where it spills into the gravel or soil. As it does, the bacteria in the effluent are destroyed by oxygen in the soil and convert into harmless compounds.

Septic tanks are designed with baffles to make the environment inside airtight and calm. Sewage enters slowly and without disturbing the processes going on. The bacteria, in essence, have enough time to work on the sewage before it disappears into the absorption field.

Most septic tanks are made of concrete and are squarish in design, a feature that helps ensure that the process is controlled. *See also* CESSPOOL.

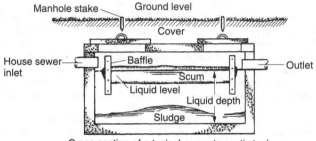

Cross section of a typical concrete septic tank

Septic tank

service entrance box The electrical box inside the home where wires from outside are connected to circuits within the house.

service entrance equipment The equipment that constitutes the heart of a home's electrical facilities. It is composed of three parts: the electric meter, the disconnect switch, and the distribution panel.

Also known as the "watt-hour meter," the meter measures the amount of electricity consumed. It may be either inside or outside the house. The meter is usually outside at newer homes. It is easier for the meter reader to reach since no one needs to be home to grant entry. In older homes, the meter is often in the basement, the idea being that the best way to protect the meter is to keep it shielded from weather.

The disconnect switch is a single switch that is rated at the same amperage as the entire system that can be used to turn the entire system off for safety purposes, such as in a fire. The switch must be housed in a metal box with an external off-on lever; it must be visible and not hidden behind something; it cannot fall closed accidentally; and it should be protected by a fuse or circuit breaker. Some utilities prefer to have the disconnet switch outside the house; this arrangement makes it unnecessary to enter a burning building to turn off the electricity.

set Hardening of MORTAR. In initial set, mortar reaches a partial hardness; in final set, full hardness.

setback Distance from the ends or sides of a lot beyond which construction may not extend. The setbacks may have been established by RESTRICTIVE COVENANTS, filed plots of subdivision, BUILDING CODES or ZONING LAWS.

sewer Pipe that receives waste products from a building. Sewers started out as overflow pipes for fishponds, but as time went on, they acquired their current meaning. Sewers are ordinarily part of a municipal waste system.

shading Dark/light appearance of ASPHALT ROOFING SHINGLES in certain spots.

This problem usually occurs with darker asphalt shingles. Because of slight differences in the manufacturing process, the shingles may be slightly different colors. When the light reflects off them, depending on the amount of light and the angle of the viewer to the roof, the roof can seem darker or lighter in spots.

There is not a great deal that can be done about this. Sometimes the cause is that the backing material used to keep shingles from sticking together transfers to the shingle face. If that is the case, normal weathering will wash the blemishes away.

shake A thick wood shingle that is cut to shape by hand. Shakes are made of rot-resistant woods such as cedar and redwood. They are actually split by hand, hence their name. *See also* CEDAR SHAKES.

shark fin A FELT sidelap or endlap on a BUILT-UP ROOF that is curled upward and looks like a shark fin.

sheathing 1. On SIDING, the plywood or other material fastened to STUD WALLS.

As on ROOFS, individual board sheathing has been largely supplanted by panelled material, so-called engineered wood, WAFERBOARD, ORIENTED STRAND BOARD, and PLYWOOD. The advantages of these panel materials for the builder are many. First, they are easier to install than individual boards. Secondly, the framework may not need BRACES if these panels are used. Using plywood makes a structure many times stiffer than using individual boards.

Normally, panels are installed vertically; but in hurricane country, they are installed horizontally, which provides greater strength, in part because they are nailed to more studs (*see* HURRICANE LOADS). Panels are spaced to allow for expansion and contraction with the gaps greater—sometimes double—than they would be if installed in areas where expansion and contraction of panels is expected.

In some areas of the country, such as the Southwest, sheathing is not used. Instead, plywood panels are nailed directly to the studs. Because the exterior side of a plywood panel is decorative in some way, they can be used as both sheathing and finished material.

2. On roofs, the material used as a base for roofing. Commonly known as DECKING.

Once, as with siding, individual boards were used for roof decking, but these have largely been replaced by plywood. Plywood sheathing greatly increases the strength of the house. Panels are installed with their face grain across the rafters, and staggered so their ends are not nailed to the same rafters.

The American Plywood Association provides tables that indicate how thick panels should be, depending on rafter spacing. Normally, ½″ material is used, either interior or exterior, depending on how much precipitation there is in the area. *See illustration on following page.*

sheet flooring *See* RESILIENT FLOORING.

shell The wood skeleton of a house with only FRAMING, SHEATHING, and DECKING in place.

shims Tapered wood strips used to fill gaps to make structures level and plumb. Shims can be made from all kinds of wood scraps, but wood shingles work well because they are naturally tapered and cut easily.

A favorite use for shims is in making kitchen cabinets level, but there are many others.

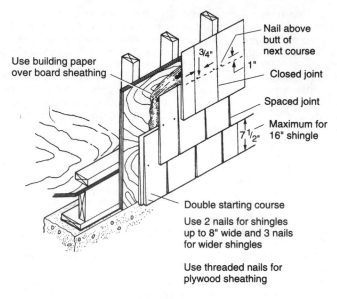

Use building paper over board sheathing

Nail above butt of next course

3/4"

1"

Closed joint

Spaced joint

Maximum for 16" shingle

7 1/2"

Double starting course

Use 2 nails for shingles up to 8" wide and 3 nails for wider shingles

Use threaded nails for plywood sheathing

Sheathing

shiplap A kind of joint.

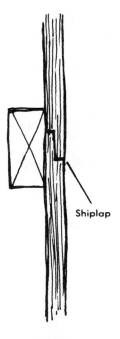

Shiplap

Shiplap

s--t Slang for JOINT COMPOUND.

shoring Bracing placed against a wall or under a beam to provide temporary support.

short circuit A malfunction in a building's electrical wiring.

shoved joints Head or vertical joints made in BLOCK by buttering the ends and then shoving them against previously placed block.

side lights Windows on the sides of a doorway.

siding Any of a variety of materials used as a final covering on the exterior walls of a building.

Like ROOFING, siding is sold by the SQUARE, 100 square feet of material. The average house utilizes 20 to 25 squares, or 2000 to 2500 square feet.

Installation of siding varies according to the material, but there are some general considerations. In most cases, if siding is applied to a previously sided structure, the old siding can remain and even serve as added insulation. If the new siding cannot be applied directly over the old one, FURRING strips can be installed to allow this to be done. BOARD INSULATION can also serve as a base, but its cost is hardly ever recouped in terms of heat savings. Its R-VALUE is low, being four, five, or less.

Aluminum siding comes in a variety of styles, textures, and colors. It is possible to get products with coatings to make them shed dirt better. Like vinyl siding, aluminum fades with time, but it can be repainted, although this would seem to defeat its low-maintenance feature. Much aluminum is 0.019″ thick, which makes it necessary to use backer board behind it. Thicker, 0.024 gauge does not require backer board. The thicker gauge is the better material.

Aluminum siding installation begins at the bottom with a starter strip and proceeds upward, the panels interlocked and nailed in place through perforated tabs in the tops. If existing siding is wavy or ridged, vertical or horizontal furring strips can be applied to achieve a level surface.

Board siding composed of long, relatively narrow wood boards in various sizes and profiles.

Board siding comes in a variety of woods. Some, such as pine, are designed to be painted; others, such as redwood and cedar, are designed to be clear-coated or left to weather to a gray color. In general, the material comes in thicknesses of ½″ to ¾″ and in widths of 4″ to 12″ NOMINAL SIZE. Some siding have a rough sawn side, the other smooth. The former is for staining—stain penetrates rough surfaces better and increases the life of the material; the latter, for painting.

Classic board siding is the bevel type CLAPBOARD, boards that are tapered in profile and that come 4″–8″ wide and with ½″ butt thickness or from 8″–10″ wide with ¾″ butt thickness.

Reverse board and batten siding is one in which narrow batten are nailed vertically to wall framing, and wider boards are nailed over these so that board edges lap battens. In this installation, a slight space is left between adjacent boards. The pattern can be simulated with plywood by cutting wide vertical grooves in the face of the ply at uniform spacing.

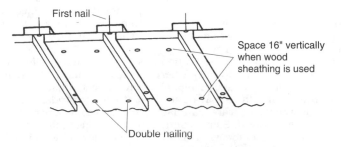

First nail

Space 16″ vertically when wood sheathing is used

Double nailing

Siding: reverse board and batten siding

Drop siding comes in several patterns and with TONGUE-AND-GROOVE or SHIPLAP edges. It is obtainable in 1″ × 6″ and 1″ × 8″ sizes and is not tapered. It is commonly used without sheathing in buildings without air conditioning and for garages. In tests conducted by the Forest Products Laboratory and the Department of Agriculture, drop siding with tongue-and-groove edges was more resistant to wind-driven rain than boards with shiplap joints.

Drop siding is installed in much the same way as clapboard siding, except for spacing and nailing. Drop siding has the same to-the-weather exposure. Actual width is normally 5¼″ for 1″ × 6″ siding and 7¼″ for 8″ siding. One or two 8d nails are used at each stud crossing, depending on width.

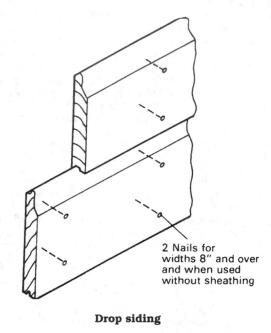

2 Nails for widths 8″ and over and when used without sheathing

Drop siding

Hardboard siding comes in large panels or board form and in a variety of textures and qualities. A good indication of the quality is the warranty: some manufacturers offer five year warranties, some fifteen. For the warranty to be in effect, the siding must be painted within a certain period of time. Unfinished hardboard has a short life.

Plywood siding panels are normally 4′ × 8′ and come in varied thicknesses (¹¹⁄₃₂″, ⅜″, ½″, and ¹⁵⁄₃₂″); with facings or veneers including redwood, cedar, Douglas Fir, or southern pine, to name a few; and with finishes from rough-sawn to textured. The finish may be stained, or painted if it is lower quality.

The American Plywood Association has established a grading system for plywood panels which indicates where they can be used, the span rating, and the nailing schedule. Some plywood panels must be secured to SHEATHING while others may be secured directly to studs.

Some plywood panels are manufactured overseas, where quality control may not be guaranteed. Some manufacturers will use interior glue for the panels to be used outdoors. It is best to look for APA-stamped panels, which also contain grade ratings information. *See* BACKSTAMP.

Steel panel siding, manufactured to simulate wood siding, is available in the same sizes as aluminum siding but is sturdier. Steel and aluminum siding are installed like other long sidings. A starter strip is secured at the bottom of the wall with nails that go through slots in the tops. Steel is strong, but if it is scratched and not touched up with paint, it can rust.

Vinyl siding comes in lengths of around 12″ and in various thickness or gauges, color, and texture. It is generally about 8″ wide and simulates clapboard; it also comes in forms that can be installed vertically.

Vinyl does not chip, dent, or scratch easily. Even when it does, it hardly matters because the color goes clear through. In cold weather, vinyl siding can become brittle and break if it is hit with a hard object, but repair is relatively easy. The bad panel is simply replaced. Vinyl also has a shallower texture than aluminum and fades more readily.

There is no specific grading system for vinyl. Rather, the siding is divided into three groups defined by color, gauge, profile (shapes), texture, and sheen.

Builder's grade, the lowest quality group, is available in a limited number of colors, chiefly white, gray, cream, and clay. It has a gauge of around 0.040″ and one or two styles or profiles. Builder's grade vinyl siding generally has a high-gloss finish. In the midrange of quality products is a group with a broad range of colors (seven to twelve). It has a richer, thicker gauge (0.0423 to 0.044), and it comes in a broader selection of profiles. These have a lower gloss than the builder's grade. The premium grade of vinyl siding is even thicker, has a lower sheen, and can have special designs and textures.

Gauge is important in vinyl siding. In thinner gauges, the vinyl will reflect the surface below, giving new siding a wavy look. Even the heavier gauges may not hide a ridged substratum, meaning FURRING STRIPS or backer board will have to be used.

Vinyl siding is not difficult to install; it is hung rather than nailed to the house. The vinyl panels have slots in the top through which nails are driven, though not so tightly the vinyl cannot expand and contract. If it cannot move, the vinyl may buckle.

Installation begins at the bottom of the house. As the panels are nailed in place, the one above hooks into the one below. The ends of the panels slip into various types of formed molding. The vinyl is easily cut with a circular saw equipped with a bruit designed to cut plastic.

Vinyl siding is popular, but it is not loved by all. In some communities with homes that have been declared historic landmarks, it is against the law to install vinyl, the feeling being that it just does not replicate the real thing. *See illustration on following page.*

See also CEDAR SHAKES (SIDING), CEDAR SHINGLES (SIDING), *and* CLAPBOARD.

sill 1. The horizontal board(s) secured to the top of a FOUNDATION wall with ANCHOR BOLTS. Also known as a sill plate. **2.** The flat shelf-like part on the exterior of a window. *See illustration on following page.*

sill block Solid concrete masonry unit used for sills or openings.

sink "Sink" specifically refers to a kitchen sink, but the term is commonly used to also describe LAVATORIES, the bathroom sink.

skim coat A thin coat of JOINT COMPOUND over a rough wall to smooth it.

Some walls are in such poor condition that patching does not yield a finished-wall look. After the wall has been patched as much as possible, the mechanic will use a broad trowel to apply repetitive thin coats of joint compound to bring it to a seamless levelness. Joint compound sticks tenaciously, so thin coats are possible and one need not be concerned about the material falling off.

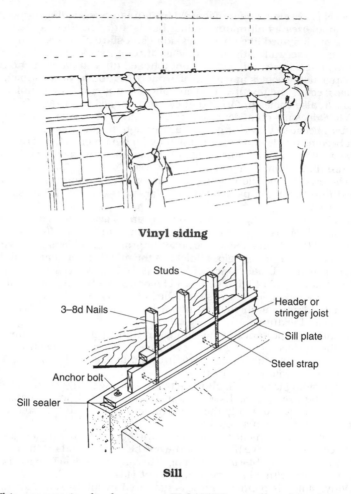

Vinyl siding

Studs

3–8d Nails

Header or
stringer joist

Sill plate

Steel strap

Anchor bolt

Sill sealer

Sill

This process is also known as FLOATING a wall and is also used to resurface areas that are partially damaged or have wallpaper on them that cannot be removed without greatly damaging the wall.

skylight A window in a roof. Skylights may be plastic or glass. Plastic skylights are rounded; glass skylights are always flat. Skylights may be either fixed or movable.

Skylights were an outgrowth of the gun turrets in American aircraft. In a sense, someone took them off the bombers and put them into the roof.

Skylights have grown in popularity in recent years, and it is easy to see why. They can flood a room with light, provide solar heat, and help to reduce energy bills. If they open, they can be good for ventilation in any room, particularly in the bathroom, where they can be opened to bleed off moisture-laden air.

Acrylic skylights come with either a single or a double dome made of a two-layer sandwich of plastic. The double dome air space may be sealed or unsealed. Unsealed double domes can be problematic; moisture can get in and create CONDENSATION.

Acrylic skylights come in a variety of colors. White is the most common, but white, gray, and bronze are also available. The clearer the material, the more light comes in. A clear skylight lets about 90 percent of the light in, but translucent gray lets only 25 percent in.

Glass for a skylight should be TEMPERED. LOW E GLASS is also available.

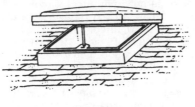

Skylight

slab 1. A platform made of concrete on which house framing members are erected. Slabs are only used in basementless houses. Also known as slab construction. *See also* FOUNDATION. **2.** A slab of concrete used as a patio.

slaking Adding water to a particular material, such as LIME.

slate Thin, hard natural stone. *See* ROOFING.

sleeper Members installed on concrete or other floors to support flooring. They are also known as "screeds." Their original name was "ground joists."

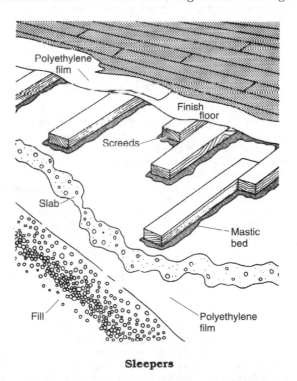

Sleepers

slider WINDOW with sections that slide back and forth to open and close.

sliding door Door with sections that slide back and forth. *See* PATIO DOOR.

slippage Lateral movements of a built-up roof. Such roofing mainly moves out of a position on a sloped roof.

slope One of the ways the steepness of a roof is measured. Slope is calculated after calculating other factors, such as rise, run, and span. The rise of a roof is its height, the distance from the top of the wall. The run is the distance between the outside edge of the roof to an imaginary line that bisects the ridge board. The slope is how many feet a roof rises in relationship to its run. So, a roof that rises 6′ in a 12′ run has a slope of 6 in 12. One that rises 4′ in 10′ has a slope of 4 in 10, and so forth. The kind of roofing to be used is partially determined by a roof's slope Also known as PITCH.

slump block Concrete BLOCK produced so it sags or slumps before hardening.

slushed joints Head joints filled in with mortar after the unit is laid by throwing mortar in with the edge of a trowel.

smoke chamber Enlarged area in a fireplace between the bottom of the flue and the top of the throat. It is designed to prevent downdrafts. *See also* CHIMNEY.

soffit Horizontal portion of an eave. *See also* CORNICE.

soffit ventilator Vent opening in the soffit that allows heated air from an attic to bleed off.

soft water Water with few minerals, the opposite of HARD WATER. Water passing through a pipe coats it thinly with a protective layer of minerals. For this reason, soft water, which has few minerals, can be problematic for copper pipe. *See also* BUILDING CODE.

softwood Lumber made from leaf-bearing trees.

Softwoods—pine, fir, hemlock, spruce, and cedar, to name a few—are the main materials used in building for which beauty is not a criteria. Softwoods are commonly used for house framing, for example. On the other hand, there are softwoods, such as cedar, which are quite good looking.

Softwoods are broken down into three categories according to size. Timbers are biggest, having a NOMINAL thickness of 5″ or more. Planks, which range from 2″–5″ thick nominally, are in the DIMENSION LUMBER category. Finally are BOARDS, which include any members that are nominally less than 2″ thick.

Boards with a nominal thickness of 1″ are the most common. They are sometimes called ONE BY as well as three quarter. (Their actual thickness has shrunk over the last 25 years or so.) Other terms used for boards include five quarter and HALF BY; the latter refers to nominal 5/4″ boards.

Softwoods are also graded according to water content, appearance, and manufacturing method. Since regional associations do this with different grading stamps, it can be as easy to straighten out as a plate of spaghetti. And then there are categories within boards that relate to its knottiness. The more knots, the lower the quality and appearance. Lumberyards may carry three grades: common, which has quite a few knots; select which has fewer; and clear, which has none.

Perhaps the best way to buy boards is simply to examine the boards, not only for knots but also for straightness, cupping, and the like by holding the board up at one end and sighting down it.

Dimension lumber has the same grading system, which makes things easier, but hardwoods have yet another system, which can be deceiving. Essentially, the proof is in the pudding. If the lumber looks good, then it is.

solar heating Heating a building with a system designed to use the sun's heat.

solder 1. In plumbing, the process of connecting copper pipes with hot solder. It is more commonly known as sweating the pipe. **2.** In electrical installations, the process of joining or attaching wires with hot solder.

solderless connectors In electricity, hat-shaped deices used to connect wires without soldering.

Solderless connectors are color-coded according to the size of the wire they accommodate. Attachment is the same, regardless of coloring. The ends of the bared wires are twisted together clockwise, and pushed into the connector. Inside the connector is a threaded portion; as the conductor is twisted, the wire wraps into it and locks into place.

Solderless connector

solid top Concrete unit that has a solid top for use as a bearing surface.
soup Hot lead used in sealing CAST IRON joints. *See* DWV SYSTEM.
spall Flaking off of the surface of concrete.
span The horizontal distance from EAVE to eave.
specs Specifications. In building, these refer to the products and the way something will be built.
splash block A formed piece of concrete designed to break up and distribute water running out of a downspout.

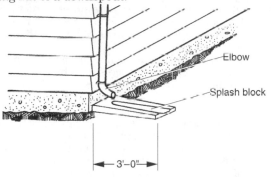

Splash block

spotting the nails Covering exposed nail heads in DRYWALL installation with dabs of JOINT COMPOUND.
spread footing *See* FOOTING.
square *See* ROOFING *or* SIDING.
stabbing Roughening brickwork to provide a toothed surface for PLASTER.
stack *See* DWV SYSTEM.
stain 1. Coating the main characteristic of which is that it colors and penetrates the wood. There are a number of differences between paint and stain, but the chief one is that paint is a film and stain is not; the latter works by penetrating into the pores of the wood. As such, when it is used outside the house, it cannot peel because moisture passes harmlessly through it.
　　Stain is available in a variety of forms for both interior and exterior use. For the outside it is available in clears, transparents, semi-transparents, opaques, and solids. Some stains have inhibitors to help the wood resist the rays of the sun, which otherwise would kill wood cells and turn wood gray. (Some clears do not have inhibitors.) Of course this does happen

eventually, sooner rather than later for a deck that takes a direct hit from the sun.

Although clear stains are supposed to be clear, they have some color in them that varies from manufacturer to manufacturer.

Transparent stains are those that are clear coatings tinted with a small amount of color. They are also known as toners because they tone or color the wood slightly. The grain, texture, and color of the wood beneath show through.

Containing even more color are the so-called semi-transparent stains. These have enough color in them to obscure most of the wood color, yet they allow texture and grain to show through.

Next up the line are opaque stains, which contain even more color than semi-transparents. They allow the texture of the wood to show through clearly.

Finally, there is solid stain, which could be considered a paint. It only partially penetrates the wood. It is essentially a film that lies on the surface, and it is susceptible to peeling if water gets behind it.

At least one company makes weathering stain to give new wood an old-wood look in a short period of time.

Interior stains work the same way, as suggested earlier, as exterior stains do: they penetrate the wood. There are some exceptions, but interior staining is a two-step process: the wood is stained and the stain is allowed to cure, and then a clear coating—a POLYURETHANE—is applied to give the stain protection. In other words, the stain colors the wood but does not afford any protection itself.

Interior stain is usually oil-based, making paint thinner necessary for cleanup. Lately, a series of water-based stains have been introduced, leaving clean up and thinning to soap and water.

Application of interior and exterior stain is essentially the same: the product is applied so it is "driven" into the wood. On exterior stains, a brush is still the best applicator because it can work the products deeply into the wood pores. For interior application, you can use a brush to apply the stain to the wood, and use cheesecloth or a staining pad to wipe off excess and work the stain in.

2. Marring of wood. There are a variety of stains that can mar wood, including water and rust.

Water-soluble extractives are natural compounds such as tannins, which are found in cedar and redwood and are in large part responsible for the beauty and stability of these woods. However, a problem occurs when water penetrates the wood and these water-soluble compounds rise to the surface. If the concentration is great enough, they can discolor the surface, a process known as tannin staining or bleeding. The stain may occur in several ways. The chemicals may darken the wood greatly, the stains may show up around pressure points such as nails, or the compounds may flow down because of gravity. On a structure's windy side, for example, one may see a vertical as well as horizontal stain lines.

Pine and Douglas fir can also have staining problems, and such stains can occur even if a surface is painted. Water can penetrate paints that are porous, dissolve the extractives in the wood, and draw them to the surface. The same thing can occur if there are gaps in the siding, or from faulty roof drainage or gutters.

The water source can also be inside the house. There can be too much water vapor created from a variety of sources from appliances to showers. This moisture is carried throughout the walls, where it seeps into the siding from the back, and dissolves the extractives in the wood. These run down the siding, leading not only to discoloration but also to coating failure.

Iron is the source of several kinds of stains with a variety of causes. Rust is one type of iron stain. When standard ferrous nails are used on exterior siding and painted, a red-brown discoloration may occur around the head. A primer will seal off the rust spots; or the nailheads can be driven beneath the surface, covered with wood putty, and repainted.

To prevent the problem in the future aluminum, stainless or high-quality galvanized nails should be used. The high quality is important because otherwise the galvanizing may chip off as the nails are driven, exposing raw steel.

Rust stains can also occur when standard ferrous nails come in contact with coatings such as solid color and semi-transparent stains. They also result from screws and other metal objects that are subject to corrosion and leaching.

Iron stains may also be the result of a chemical reaction with wood. Iron reacts to certain extractives in the wood to create a blue-black discoloration. Ferrous nails are the most common reason for chemical staining, but stains can also result from cleaning the surface with steel wool or wire brushes.

Oxalic acid will remove the blue-black discoloration provided it is not already sealed beneath the finish coating. To do this, the surface should be given several applications of a solution containing at least one pound of oxalic acid per gallon of water, preferably hot. After the stains disappear, the surface should be thoroughly washed with warm water to remove the oxalic acid residue. If all sources of discoloration are not removed, the problem can recur.

See also MILDEW.

The myth of letting wood weather

For years, it was felt that the best way to handle exterior wood was to let it weather a few months or so prior to staining or painting. Popular belief had it that the weathered wood would be better able to absorb the stain or paint and stick better.

Research by the Forest Products Laboratory, an arm of the US Department of Agriculture, has found this to be false. In fact, wood starts to deteriorate immediately, making it important to get some sort of protective coating on it as soon as possible. The main reason is water. When wood gets wet it expands; and when it dries, it contracts. Over a relatively short period of time, the wood can deteriorate and crack from the cycle of wetting and drying. Moreover, wet, unprotected wood is an invitation to mildew, and the sun damages it as well, destroying wood cells. In sum, it is estimated by the FPL that wood left unprotected will have its life shortened by twenty percent.

The myth seems particularly credible for PRESSURE-TREATED LUMBER, the idea being that this can withstand weather well. In fact, pressure-treated wood is wood; the chemicals in it protect it from fungi and insects, not water and sun. Pressure-treated wood needs to be protected as soon as possible. It does not matter that it is fresh wood. As soon as the surface is dry, it can be coated.

Blocking the sun

The invisible ultraviolet radiation in sunlight is the natural enemy of wood. It attacks the lignens, the connective tissue of wood, graying and eroding it over a period of time.

This cannot be stopped, but the presence of pigment in stains greatly slows it down. (Indeed, paint would protect it completely.) Clear coatings may also contain organic or inorganic chemicals that guard against UV radiation.

Organic inhibitors are carbon-rich chemicals that cling to the wood and work by chemically absorbing heat on a molecular level. These inhibitors sacrifice themselves in the process, so the UV protection declines over time and new material has to be applied.

Inorganic inhibitors block the UV rays. They are made of iron oxides ground to an ultra-fine consistency. The finish is clear to the human eye, but opaque in the UV range, creating a layer of protection.

One can tell by eye if organic or inorganic inhibitors are used in clears. Clears using organic inhibitors will appear water-clear, while those using inorganic inhibitors will have slight pink or reddish cast.

stainless Made with nickel and chrome, this shiny metal is generally weatherproof and strong, but stainless comes in degrees of quality.

stair treads *See* TREADS.

staking out Marking off a building site for excavating with stakes and string. *See* BATTER BOARD *and* EXCAVATION.

starter course *See* SIDING.

stepped footing FOOTING used when a lot slopes. A stepped footing resembles a series of steps and are located on solid earth below the level of the FROST LINE to guard against heaving.

Stepped footings should follow certain dimensions. Considered as a step, the vertical or RISER portion of the step should be at least 6″ high but not more than three quarters of the footing width nor more than 2′ high. On steep slopes, more than one step may be required. For this, the lower portion of the footing is poured, and then the upper. The two are tied together with reinforcing rod.

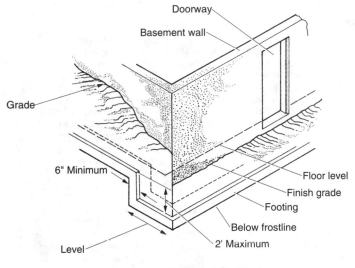

Stepped footing

stick built Frame house built with wood members BALLOON or PLATFORM FRAMING. The term likely comes from the appearance of the framing, which is thin in comparison to the BRACED-FRAME construction that proceeded it.

stiff mortar MORTAR low on water.

stile *See* DOOR.

stinkpipe A building's soil stack, into which all fixtures empty. The term describes the gases, such as methane, that are generated from the decomposition of sewage. The top of the pipe usually extends 3′ or 4′ above the roof line. *See also* DWV SYSTEM.

stirrup Support that resembles a saddle stirrup used to support the end of a JOIST or some other heavy member of a structure. Stirrups are also known as "hangers" and may be suspended from a GIRDER as well as a wall.

stone Name for a variety of stones used for building purposes.

The stone available for building depends on the area one lives in. It might be simple to get a wide array of slates in Vermont, but not in Arizona.

Stone is usually stored in a yard according to size and type. A well-stocked yard will also carry large stones for MANTELS and ARCHES. Stone is usually available in three or four finishes or grades, as indicated later.

Laying a hundred stones four or five cubic feet requires mortar. Stones may be joined by bonding with type M Mortar or left unbonded. Such unmortared or dry-laid stone is usually used to make retaining walls or as decorative accents in landscaping projects.

Stone, like other MASONRY products, comes with different weathering characteristics. Closely questioning the stone dealer should reveal which arc best. Stones come dressed, semi-dressed, undressed, rounded, and veneer.

- **Dressed stones.** Custom-trimmed to the exact dimensions a buyer wants. Such stones are very expensive.
- **Semi-dressed.** Cut on all sides to the approximate shape needed; trimmed further before actual use.
- **Undressed.** Stone as it is mined from the quarry, unsized and uncut.
- **Rounded.** Stone that is gathered from river beds.
- **Veneer.** Any of a variety of stones, 2″–3″ thick with flat-cut backs. Veneer stone is installed on steel stud walls with a facing of expanded metal lath. The stones are laid in place after type M mortar has been troweled onto the lath. The stones must be wet first to ensure that they will not take moisture from the mortar and weaken it.

stool *See* WINDOW.

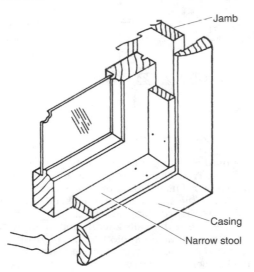

Jamb

Casing

Narrow stool

Stool

stoop Structure consisting of stairs leading to a flat landing. "Stoop" comes from a Dutch word meaning step.

stop 1. Thin molding installed on a door to limit the distance it can travel, or a cut in a door that does the same thing. **2.** Molding installed on a window frame to keep the window's action perfectly vertical.

storm door DOOR installed outside an entry door. A storm door is designed to help insulate against cold wind and weather.

storm window Protective window installed outside regular window.

stor pole Marked pole for measuring masonry units during the building of a wall.

stove bolt This bolt is threaded along its entire length and has a slotted head and has a nut. It is used to fasten metal items but works well on wood as well.

S trap In plumbing, a TRAP that looks like the letter S. This type of trap is used where the WASTE pipe feeds into the floor rather than the wall like a P trap does. S TRAPS are banned in most sections of the country because they can compromise themselves by creating a siphoning effect, pulling water out of the trap.

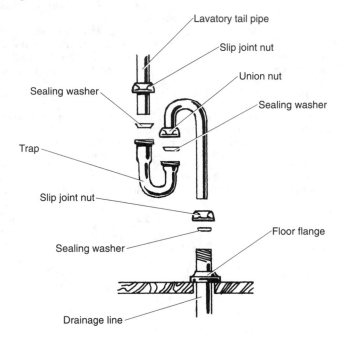

S trap

strapping 1. Iron starts used to bind framing members together in areas subject to great wind stresses. Typically, the straps are nailed to and join STUDS and box JOISTS. **2.** In areas of high wind, iron straps used to tie MASONRY walls to joists. In this case, the straps are often embedded in the masonry. **3.** In CONCRETE BLOCK construction, strapping used to tie the LOAD-BEARING WALLS to joists to give the walls lateral support. **4.** Strapping used to tie wall STUDS to rafters in areas subject to the high wind and earthquakes (*see* WIND LOAD *and* EARTHQUAKE LOAD). In these situations, nails may not be enough to hold a roof on. *See illustration on following page. See also* FRAMING FASTENER.

strawberry A small blister or bubble in an asphalt roof.

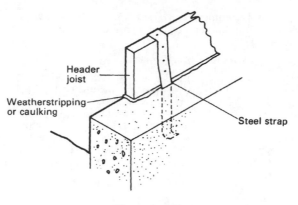

Strapping ties

stretcher *See* BRICK.
stretcher block Concrete block laid with blocks parallel to the face of the wall.

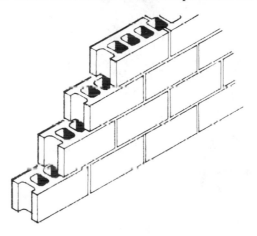

Stretcher block

string Laying mortar across a number of units at one time so a group of block can be laid.

stringer Slanting board on a staircase that TREADS and RISERS are let into.

strip flooring Narrow (3½" or less) strips of wood with tongue-and-groove edges. Strip flooring may be anything from 1"–3" wide and vary in thickness from ⅜" to ²⁵⁄₃₂" thick.

This is often called matched flooring because of the way it fits together. One end of a strip has a tongue on one side and a groove on the other. Tongues are nailed to the floor, the grooved portion is slipped over it, and that piece is nailed.

Some flooring is made from SOFTWOOD; but certain types of HARD-WOOD predominate, chiefly oak, but also maple, birch, pecan, and walnut.

Different woods are graded differently; but by far the popular wood, used for over 90 percent of wood floors, is oak. Different woods are graded different ways, but essentially it is a question of looks. Oak, for example, has four basic grades: clear, which has almost no imperfections such as knots; select, which is almost all clear but has some knots and color imperfections; and Common No. 1 and No. 2, which have natural imperfections. Some people prefer the last two grades because they want that particular look.

> ### Spiked heels: Hardwood flooring's enemy
>
> Experts on hardwood point out that shoes with spiked heels, even styles with wedge heels, contain a steel spike to strengthen the heel. This spike is the equivalent of a blunt ten penny nail.
>
> If the leather or plastic cap wears down or off as frequently happens, the nails holding in the center spike can become exposed and cause surprising damage. It is estimated that a two-ton car exerts 28 to 30 pounds per square inch of pressure, a full-grown elephant 50 to 100 PSI, and a 125-pound woman as much as 2000 PSI when taking a normal step in spiked heels.
>
> The solution, say experts, is to check heels for wear frequently and get them repaired if necessary.

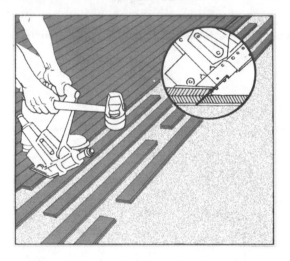

Strip flooring

struck joint MORTAR joint finished with a trowel but not tooled.

stucco Cementitious material applied to the exterior of a building. Also known as a plaster finish.

Stucco is made from PORTLAND CEMENT, LIME, sand, and water. It is applied to exterior walls like PLASTER.

Stucco is designed not only to be beautiful, but also to protect against weather. Normally two coats are applied, followed by a sealing coat of paint. Applied properly, stucco—as shown on homes throughout the world—can last for a hundred years. *See illustration on following page.*

stud walls Walls made with STUDS.

In America today most of the wall framing is conducted with stud walls, either metal or wood. Wood is far more common, but metal is gaining.

subfloor The first flooring, the material nailed to JOISTS.

substrate The base material on top of which other material is installed.

switch A device to cut off or allow the flow of electrical CURRENT. When the switch is flipped or turned to the OFF position, inner metal contacts separate. Flipping the switch back on remakes the contact and allows the cur-

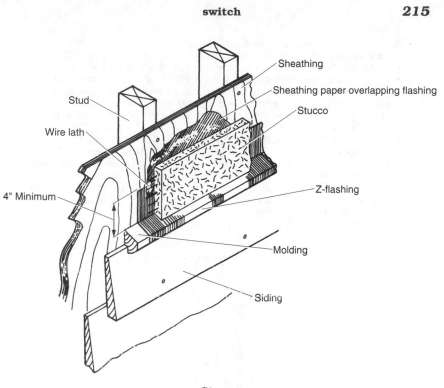

Stud

Wire lath

4" Minimum

Sheathing

Sheathing paper overlapping flashing

Stucco

Z-flashing

Molding

Siding

Stucco

rent to flow again. There is a wide variety of switches available, but most common are house switches.

Many of the things that can be said about RECEPTACLES can also be said about switches. Like receptacles, switches are designed to handle specific electrical LOADS. Commonly, switches come in fifteen- and twenty-amp sizes. This means that the contacts in them can handle up to 15 or 20 amps of current without being overloaded.

Twenty-amp switches are known as "spec" or specification grade. They not only carry higher amperage; they are better made. If a switch is used frequently the use of a spec grade is justified.

House switches are often referred to as single pole and double pole. The former means that they interrupt the "HOT" side of the circuit. The latter means that either both the hot and neutral sides of one circuit, or the hot side of two circuits, is interrupted.

Most switches have screw terminals—brass, silver, or green-colored screws—on which wires are wrapped around clockwise to make electrical connections. They also have holes in the back or side for clamp connections. Wires can be pushed into these holes and clamped in place, rather than wrapped around screws. Having clamps and screws creates more capacity for accommodating the wiring connections. The holes are an integral part of the connections to screw terminals.

Switches have three kinds of on/off action, ranging from loud to silent. Some have a distinctive click; a mercury switch is silent.

Like receptacles, switches have metal stampings on top with tapped screw holes for ready attachment to a box. They also come equipped with ears so you can install the switch flush with the wall material.

A variety of switches are available, as follows:
- **Low voltage.** Switch used with wiring that carries twelve volts or less.
- **Outdoor.** Switch constructed for use out-of-doors without damage.
- **Single pole.** Rectangular box with a toggle switch in front and one brass screw on either side or both screws on one side. Single-pole switches are used to control one light or one group of lights from a single point in the room, typically from the room entrance. It controls one 110 VOLT circuit.
- **Three-way.** Switch that can control a light from two different directions. A three-way switch could be installed at the top of the stairs to a basement and another at the door downstairs that leads outside. After turning the lights on before you go down the stairs, you could turn them off as you leave from the other switch. A three-way switch looks like a single-pole switch except there are two brass screws and one copper one.
- **Four-way.** Switch that can control lights from three different directions.

Single pole switch

Duplex switch

Three-way switch

tab *See* ROOFING.

tail beam Any joist or other timber that is flush with the header.

tannin Naturally occurring compound in all species of wood that can cause stain problems. *See* STAIN.

tar Black or brown bituminous material, liquid, or semi-liquid in consistency. Tar is mostly made from bitumens gotten as condensates from coal, oil-shale, petroleum wood, or other organic material.

tarmac Name of a company and a type of road material made of tar and macadam, a paving material made of compacted small stones.

> ### How tarmac came to be
>
> Tarmac came about as the result of an accident around the turn of the century. A barrel of tar had fallen on the road, and a man named E. Purnell Hooley spotted it. Apparently some workers had tried to cover it up with macadam, and as Hooley is quoted by author David Owen in his book *The Walls Around Us*, the resulting composite "had properties far superior to the road surface of the day."
>
> Hooley made some refinements to the material and patented it under the name Tarmac. Known as the TarMacadam Syndicate when it was formed in 1903, the company is today known as Tarmac PLC.

tarp Short for tarpaulin, which can be used as a drop cloth or to cover an entire house, depending on size.

tearoff Removing roofing down to the deck, or wood base. Also known as "ripoff," this is the most common term for stripping a roof of the roof covering.

It is a very messy job. First, heavy tarpaulins are tied to and draped down off the gutters (assuming they are coming down) to cover any vegetation by the house. The tarpaulins extend a few feet out to catch any broken pieces of shingles. Also parked as close as possible to the house will be a dumpster to toss the debris into.

The other alternative is to lean halves of an extension ladder against the house and tie sheets of plywood to them, forming a protective half-tent.

The chief tool in removing shingles is the "shingle eater," which looks like a long-handled shovel with serrated teeth. The job begins at the top of the roof near the RIDGE. The shingle eater is driven under the shingles until it contacts nails, and pried upward, lifting nails and shingles out at the same time. At the same time, the building paper underneath, which is also nailed on, is taken off.

In rare cases, the decking has to be removed. Sometimes there is some damage, but rarely is it major.

tee In plumbing, a T-shaped FITTING with one vertical section and two straight sections that come off it to form the top of the T. It allows water or waste to travel in three different directions.

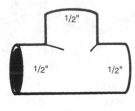

Tee

tee wye In plumbing, a FITTING that, as the name suggests, is a blend of a T and a Y.

telegraphing Show-through on a smooth overlaid plywood panel surface of underlying grain or defects.

tempered glass *See* GLASS.

template A pattern used to make an outline for cutting something.

tenon *See* MORTISE AND TENON.

tensile The strength of an item when lateral pressure is put on it.

terminal The screw on a receptacle or switch.

Terminal

termite shield Sheet metal installed on top of a foundation wall to protect the wood above against termite attack. *See illustration on following page.*

terrazzo flooring Small compacted stones of various colors embedded in concrete. The floor is normally given a high polish.

textured finish Having a rough or worked finish. PLYWOOD, STUCCO, CONCRETE, and a variety of other materials may have a rough or textured finish. A textured finish is also achieved by applying texture paint.

texture 1-11 American Plywood Association name for a SIDING PANEL with ⅜″ grooves typically spaced four or eight inches on centers. They connect with SHIPLAPPED edges.

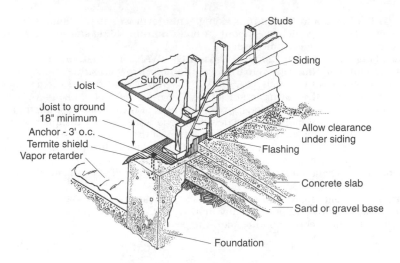

Termite shield

thermal shock Stress produced in a material's resistance produced by temperature changes in a wood membrane.

thermostat A device for setting the heat level in a building.

thin set A type of glue used to install CERAMIC TILE. *See also* ADHESIVE.

three ply *See* PLYWOOD.

threshold Piece of wood or other material directly under a door when it is closed.

through stone *See* BONDSTONE.

tie Metal strip used to connect masonry units to each other or to other items.

tie beam Member that is secured to lower ends of rafters to prevent them from pulling apart.

tile hanging Trade lingo for installing CERAMIC wall TILE.

timber framing Method of framing a house that relies on heavy timbers and elaborate joinery. Also called POST-AND-GIRT FRAMING.

Timber framing, also known as post-and-girt framing, dates back to medieval times. It was the most common framing until BALLOON FRAMING supplanted it in the early 1800s, although it is still practiced in a limited way today. The basic methods are still used in constructing large buildings, including skyscrapers.

Like PLATFORM and BALLOON FRAMING, timber framing has a SILL, but it is composed of heavy beams laid around the edge of a masonry foundation and joined together by MORTISE AND TENON JOINTS or some other fancy joint. For further strength, pegs are driven in though holes made in the joints.

Large joists, called GIRTS (they might just be tree-trunks shaved flat on top so flooring would lie flat) were then laid across the sill. A large post, often a foot square and two stories high, was tenoned into the sill. These were tied together at the first and second floor with plate members, and then the rafters were leaned together, notched into the plate, and notched and pinned together at top. Unlike platform or balloon framing, there was no ridge or ridgepole in early timber framing. To facilitate nailing of siding, relatively thin STUDS were placed between the posts, as in platform building.

Erecting such a structure required plenty of muscle power and time. In contrast, a single carpenter using platform framing, might erect the shell of a small house in a few days.

tin 1. Trade lingo for aluminum siding. This term was made famous by the movie *Tin Men*, which was about early aluminum siding salesmen. **2.** Procedure used in electronics work.

tits Slang for projecting heads of nails in DRYWALL installation.

tobacco juicing Name given to the formation of a brown residue that runs out of asphalt products such as ASPHALT ROOFING SHINGLES. Tobacco juicing occurs as the normal result of weathering. The root cause is a certain chain of exposure conditions: prolonged lack of rain, intense sunlight, and accumulation of moisture and dew. Tobacco juice usually disappears after the first rainy season following installation of the roofing. In some cases, it is necessary to take other steps, including regularly hosing down the roof or applying protective coatings.

toenailing Driving nails in at an angle at the bottom of wood members.

toilet *See* WATER CLOSET.

tongue Part of a board that sits in a recess. *See* TONGUE-AND-GROOVE.

tongue-and-groove In general, any item with a projecting piece, the tongue, that fits into a corresponding slot, the groove. *See* STRIP FLOORING.

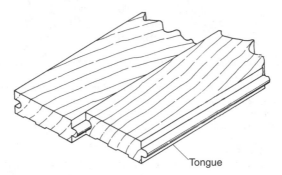

Tongue

Tongue-and-groove

tooled joints MORTAR that has been pressed and smoothed in place to maximize weather resistance.

tooth In painting, the capacity of a surface to help hold a painting coating on. *See also* PRIMER.

toothing Arrangement of brick so that alternate courses of brick project for better bonding with brick that follows.

top plate *See* PLATFORM FRAMING.

track lighting Surface-mounted or flush strip of track that houses a series of movable lights. Track lighting is mainly intended as an accent light to spotlight art objects, a painting, or the like. The lights, in separate housings, can be manipulated in a variety of directions. Power is usually supplied from a ceiling FIXTURE or by running a wire to a nearby RECEPTACLE.

trap U-shaped section of pipe on a fixture or pipe that traps water as a seal to keep gas and vermin from entering a building through the pipes. Traps vary in form. Under sinks and lavatories, traps are either built in or are separate pieces of pipe, either P traps, in the shape of the letter P; or S TRAPS, in the shape of the letter S. The main drain or sewer line has a large U-shaped trap at the end of it; a humped-over section in a toilet has a trap built in.

The action of a trap is simple, as can be seen when water goes down a sink drain. The water travels out the drain, down a tailpiece, into the trap, then onward to the drain line, which leads to the STACK. Even turning off the water leaves a small amount of water, called a trap seal or just seal. This seal must be at least two inches deep between the top of the trap's "crown weir" and "dip."

The key to a trap's proper function is air pressure, which exerts equal pressure on the water seal. A vent line in each branch line is open to the atmosphere to ensure that air pressure is operative. If this vent line fails—it might be clogged, for example—the air pressure becomes unequal and may lead to the water's being siphoned out, leaving a trap seal that is partially or fully compromised.

Another type of trap is the drum trap, a canister-shaped trap used for tubs and shower stalls. Drum traps range in size from three to four inches, allowing a considerably larger amount of water to act as a seal and thereby having greater protection against siphonage. No matter how much water flows through the trap—and considerable water does when a tub is drained—enough water will be retained in the trap to guarantee the seal.

Drum traps are located under the tub or shower, making it difficult to vent them directly outdoors; however, they can often be tied into other drain lines—so-called wet vents—and then drain outside to guarantee that air pressure is equalized, the key to a seal staying intact.

Access to the trap is from a cap that is screwed in. Sludge and sediment often accumulate along the interior sides of the trap, but there are kinds where drain water can be routed through to keep the sides clean.

An enormously important job

For such a simple fitting, traps do an enormously important job. Without the water seal they provide, vermin such as water bugs, silver fish, cockroaches, and even rats could get into a building.

The gases they keep out are likewise formidable. There are seven or eight of these including carbon monoxide, methane, hydrogen sulfide, sulfur dioxide, carbon dioxide, ammonia, and illuminating gas. Carbon monoxide is the most dangerous as it is odorless, but the others, even in low amounts, are explosive, asphyxiating, or just irritating. All except carbon monoxide have objectionable odors. Of course, in reality it is unlikely such gases would prove fatal or do serious damage, as they would be so dispersed.

Making sure it worked

In his book *The Walls Around Us*, author David Owen says that "in 1840 or so, [the trap] made indoor plumbing possible." He also notes that early trap designs were not reliable and allowed gas leaks into the house. Leaks were so common that you could buy a pair of sealed one-ounce vials of peppermint oil from druggists. After climbing up on the roof, pouring the peppermint oil in, and blocking the vent with a rag or paper, you would go through the house trying to detect a peppermint oil smell. If you could, the trap was not airtight.

tread Part of a step one steps on.

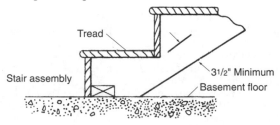

Tread

trim General term used to describe the WOODWORK of wood finishing touches inside and outside a building. *See also* MOLDING.

truss Rigid framework of triangular shapes that replace RAFTERS and ceiling JOISTS.

Roof trusses are used because they are quick and efficient. After the walls are erected, trusses are set across the TOP PLATES and nailed in place. The DECKING may then be nailed to the truss, followed by the roofing. The truss is typically capable of supporting roof and ceiling loads over long spans without intermediate support. For house construction, trusses generally span from 24 to 40 feet.

Trusses are economical because they use less material than rafter-framed roofs, and, of course, on-site labor is much reduced. Also, no LOAD-BEARING WALLS are required, allowing greater flexibility for interior planning because PARTITIONS can be placed without regard to structural requirements.

Most trusses are available with horizontal blocks called "returns" that extend from the outer end of the overhang to the exterior of the wall to which SOFFIT materials are fastened.

Trusses are normally specified by architects according to a wide variety of criteria from SPAN to PITCH.

More and more common

Today, more and more architects specify trusses for building. The essential reason is cost. The truss is prefabricated quickly in a factory without carpenters having to calculate angles and precisely cut roof members. When it arrives at the site, it is ready to be lifted into position by a crane. Additionally, because the truss is engineered a precise way, top-notch materials need not be used to produce perfectly functional members.

On the negative side, the low sloping sides of the typical truss allows far less room for storage in the attic than a normal house. If this is compensated for by adding closets downstairs, the building costs can increase geometrically. Trusses may also not be changed in any way. If they are—such as an opening cut—this can affect their structural integrity to the point of uselessness.

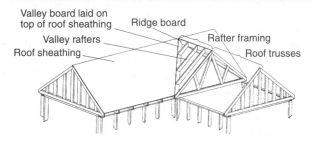

Valley board laid on top of roof sheathing
Ridge board
Valley rafters
Rafter framing
Roof sheathing
Roof trusses

Truss

tub Short for bathtub, FIXTURE used for bathing. Today there are four kinds of tubs commonly made: cast iron, enameled steel, fiberglass, and cultured stone.

Enameled cast iron is considered the "Cadillac" of tubs. It has a thick, rigid coating that is good at resisting chipping. It is also heavy; the average cast-iron tub weighs about 500 pounds.

Enameled steel is steel with an enamel coating. They weigh half of what cast iron does, but they are noisier when water runs into them and will dent and chip more easily.

Fiberglass comes in a much wider array of colors and shapes than other types of tubs, but it scratches easily. Manufacturers say the scratches can be easily removed with a Soft Scrub pad.

Cultured stone tubs are made by mixing crushed stones such as onyx granite or marble with a polyester resin and pouring the mixture into a mold with a tub shape. Appearance depends on the stone used. Onyx creates tub with a glassy look; granite looks freckled. Cultured stone has a gel coating that cannot be repaired if it is damaged.

Tubs come in a wide variety of shapes, including ones for corners. Normally, tubs come "skirted" with a frontwall, but they can also be obtained without a skirt in case a facing is going to be constructed in front of the tub. Generally, tub sides are straight, but sloping sides and configurations such as armrests are also available.

Tubs normally have bottoms that are flush with the floor, but clawfoot types, the feet of which raise the tub off the floor, are available. These relics from the Victorian age are known as "freestanding."

See also FITTINGS.

Early tubs

Bathing regularly came to America in the nineteenth century. As part of a Saturday night ritual, a large wooden tub would be dragged out in front of the stove and filled with hot water and, one by one, all family members would use it. The White House also had a tub, an elegant zinc-lined mahogany unit.

The coming of age of the tub started around 1860 with an obscure maker of cast iron pipe named Ahrens and Ott, in Louisville, Kentucky. They started to make cast iron hoppers, and when they purchased a bankrupt kettle factory, they began making tubs. Two were cast a day and one enamelled per week.

Then, in 1883, along came John Michael Kohler. Kohler was a manufacturer of cast iron agricultural implements, and he used the mold of one of the items in his line—a combination hog scalder/horse trough—as a mold for making the first tub.

Tub design limped along until 1909, when President William Howard Taft, a big man, had a tub specially designed for himself. A famous picture showed three workmen in the tub, possibly the source of the song "Rubba dub dub, three men in a tub." It certainly showed people that tubs need not be shaped like hoppers alone, and the surge in the tub's popularity began.

tuck pointing Replacing old, damaged or missing MORTAR with new material.

Tyvek Brand name for a material that wraps around a house and provides a moisture barrier. BUILDING PAPER works by keeping water out of a structure should it get past the SIDING. Tyvek, on the other hand, acts as a moisture barrier against vapor. It keeps moisture vapor out but also keeps it in, something that helps insulate. See also INSULATION.

U bolt Bolt shaped like the letter U.

underground service Refers to electrical cables that run underground from a utility's power to a house's SERVICE ENTRANCE EQUIPMENT.

For aesthetic reasons and for greater strength, most utilities like to install electrical cables out of sight under the ground. There are a number of regulations that normally apply to this. For one thing, the cable—called a service lateral—must usually be buried a minimum of 24″ below grade, encased in a metal sheath. If frost conditions warrant, it must be installed so it is not affected (*see* FROST HEAVE). Cable that moves because of a moving ground can be hazardous.

It is ordinary for the cable to be housed in conduit secured to a company pole. If this is the case, the conduit would rise to a minimum of 8′.

At some point the cable will pass through the foundation, and then rout up to the OVERHEAD SERVICE and SERVICE ENTRANCE EQUIPMENT.

underlayment The substrate or underfloor on which finished flooring material is installed. The type of underlayment used depends on what the finish flooring is. On an installation where the finished flooring and underlayment have been removed, the SUBFLOORING is often used as underlayment. When it is difficult to get the existing flooring up, the untempered type is used because it accepts nails much better than tempered type.

Plywood underlayment is installed in sections and nailed to the JOISTS using rosin-coated box NAILS or 8d cut nails. Cut nails go into the joists (*see* NAIL). The goal is to provide a smooth, level surface that is tight so the plywood does not move up and down and move the flooring itself. The plywood is cut and fit so that the ends butt on joists. If the subflooring consists of boards at right angles, the edges of the underlayment are located on the board centers. If the subflooring is over diagonally placed subflooring, it does not matter where the edges fall. It is important that underlayment joints do not fall over subfloor joists and that too many underlayment section ends do not fall on one joist.

The nailing schedule is 2"–3" apart on edge seams; the ends nailed 3"–4" apart where they butt; and 6"–8" apart through the middle of the plywood.

If a floor cannot be removed easily, ¼" untempered hardboard is used; it accepts nails much better than tempered wood. The hardboard is placed rough side up, and the nailing schedule is followed as for plywood.

Make sure that nailheads do not protrude above the surface. After a floor is nailed off, a broad knife run over the floor in several directions to make sure that none are.

Underwriter's Laboratories A not-for-profit independent organization that tests electrical products for safety.

While there are other testing organizations, UL is by far the best known. There is some confusion about what they do, and what a UL seal means when it is on a product such as a wire, switch, or receptacle. What it means is that the product has been tested by the UL and has been found safe for its intended purpose. UL may test zip cord that is used for lamps, and find it safe for its intended use—lamp wiring. It would not be listed as safe for house wiring.

Nor does a UL seal mean that the item is of high quality. For example, one switch that carries a UL listing may fail after 5000 clicks, but another does the same job 20,000 times before it dies. UL listing makes no distinction between the two; it has nothing to do with quality.

Nor does a UL seal mean that an electrical inspector will allow it in a particular installation. It may be perfectly fine for a particular use, but not for that installation.

Manufacturers submit their products to UL, which tests them in laboratories for safety. UL field representatives also visit manufacturers' factories to test products on the line, and they also buy and test products for sale in stores.

Products that have been "approved" carry a UL label.

See also BUILDING CODE.

union ell Plumbing FITTING, also called an elbow, that makes an angle between adjacent pipes. The angle is always 90 degrees unless another angle is stated.

UV Ultraviolet waves of the sun that are invisible to the eye but can damage finishing materials. *See* STAIN.

valley Intersection of sloped roof sections.

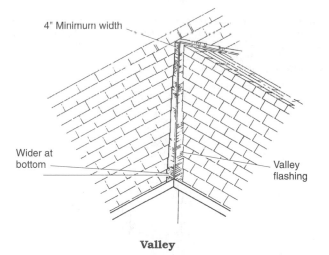

Valley

valve In plumbing, any device used to control the flow of potable water. Faucets, because they are also used for this purpose, are commonly called valves. There are all sorts of valves available that are installed on pipes:
  • **Ball.** This is a pipe-like mechanism with an angled handle. The ball valve works like a washerless faucet. Inside the valve is a teflon-coated stainless steel ball that covers and uncovers the water supply hole as needed. *See illustration on following page.*

229

Valve: ball valve

- **Check.** This resembles a coupling and keeps water from flowing in or out of the boiler too quickly.
- **Cut-off.** This valve looks something like a small round elliptical COUPLING with a handle on it. Cut-off valves come in a number of sizes, but they all control water flow. Sinks and lavatories have two stop valves: one for hot, and one for cold. Toilets have only one, which controls water supply to the toilet tank.

Valve: cut-off valve

- **Gate.** A ridged circular metal handle connected to a vertical pipe assembly. Inside the pipe assembly is a metal plate that rides up and down into a seat. Of all the valves available, a gate valve allows the greatest water flow. It is commonly the type used as the main valve in a house, and therefore must be of top quality.

Valve: gate valve

- **Globe.** A ridged, circular metal handle on a vertical stem set into a globe-like assembly. The globe valve works like a faucet. It has a stem inside, on the end of which is a washer that presses against a seat to start and stop water flow. The letter R or L is stamped on the valve to indicate which way to install it according to water flow.

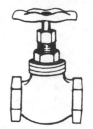

Valve: globe valve

When the valve is in the OFF position, the water may be drained out of the other side of the valve, facing vent, cap down.

• **Stop valves** have the internal mechanism of globe valves: washer and seat are replaceable.

Activity is the best policy

As with many other products, occasionally turning a valve off and on is a good way to keep it in shape. Corrosion cannot form if the handle is turned; as the rust is ground away, a fresh surface emerges.

vapor barrier 1. Material that controls moisture transmission through walls and other building elements. The vapor barrier is placed on the warm side of the wall, because that is the side that moisture-laden heat derives from. Except for chopped-up varieties, INSULATION must have a vapor barrier. **2.** Material used beneath SUBFLOORING to ensure that moisture does not penetrate subflooring or flooring. The vapor barrier recommended is normally less than one "perm" (4 mil polyethylene has a 0.08 perm rating).

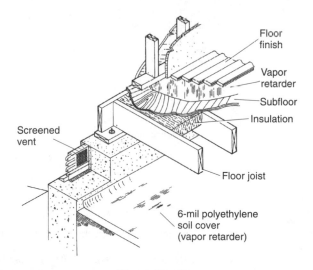

Vapor barrier

vapor migration Movement of water vapor from an area of high water pressure to one of lower pressure.

variance Permission to do a project that ordinarily would not be granted under ZONING LAWS. If a project violates local zoning laws, it is not necessarily dead. A variance can be sought, something that may require the use of an attorney.

veneer **1.** Thin sheets of handsome wood. **2.** Thin sheets of attractive stone. **3.** Thin sheets of wood bonded together.

veneer grade Standard grade designations of softwood veneer used by plywood panel manufacturers. *See also* PLYWOOD.

veneer wall MASONRY wall that is composed of thin units of masonry either attached to metal or wood studs or on a backup wall but not carrying a structural load. One can consider a veneer wall to be facing.

Veneer walls provide protection, but they also have their own distinctive look, depending on the material used as a veneer—BRICK, STONE, SLUMP BLOCK, split block, etc.

Veneer walls resemble CAVITY WALLS, but there are important differences. Cavity walls consist of two walls, both of which resist WIND and carry loads. A veneer wall is not under such pressure.

Like other MASONRY walls, flashing is used and there are WEEP HOLES in the veneer.

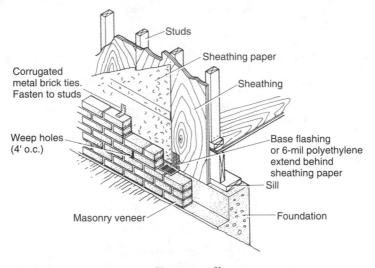

Veneer wall

vent Opening that provides an inlet or an outlet for the flow of air.

There are several kinds of roof vents, including RIDGE, SOFFIT, and GABLE end.

vergeboard *See* BARGEBOARD.

verminulate **1.** Aggregate used in lightweight insulating concrete. **2.** Material used as loose INSULATION.

vitreous china A material commonly used to make WATER CLOSETS and LAVATORIES.

This material is used extensively in the manufacture of these fixtures because it is impervious to water (vitreous means glass-like). It is a pottery product. A combination of products such as flint and feldspars are mixed with water and poured into a PLASTER OF PARIS mold. Removed from the mold and allowed to dry, the china is inspected for defects. Afterwards, the china is glazed and baked in a kiln for 24 hours at temperatures up to 2500 degrees.

People who manufacture such fixtures say that the key is to use high-quality clay. As the manager of an Eljer plant said, "If you don't start out with high quality materials, you won't end up with high quality products no matter how much inspecting or analyzing you do in-between."

Although it is highly resistant to stains, vitreous china is relatively brittle and has to be carefully handled or it can be cracked.

void 1. An open joint in a plywood panel. **2.** The CORE or empty space in a BLOCK.

voltage Term used to describe electrical pressure. *See also* AMPERE.

voltage drop The lessening of electrical pressure. This normally occurs when the device drawing the power is at a remote location from the power source. Like water, current loses its power over long distances.

waferboard A type of structural flakeboard made of compressed, wafer-like wood particles or flakes (as opposed to strands) bonded together with phenol resin. Waferboard is a relatively new material.

wainscoting General term for the material applied to the lower three to four feet of an INTERIOR wall. It is usually both functional and decorative.

In dining rooms and studies, the wainscot is usually about the height of a chairback to protect the wall when chairs are pushed against it. Traditional height is 36″, but 32″ is often used to make optimal use of 4′ x 8′ panels. A MOLDING or cap is applied at the top of the wainscot and is sometimes referred to as CHAIR RAIL.

Wainscoting is also used in wet areas such as bathrooms or room with pools or hot tubs. Here, CERAMIC TILE is usually used.

waler Horizontal timbers used to brace concrete form sections. *See illustration on the following page.*

wall Vertical component of a building of a structure helps enclose the structure and consists of framing, sheathing, and interior and exterior surfacing.

wallboard *See* DRYWALL.

wall plate 1. The cover over the electrical outlet or switch on a wall. **2.** Top horizontal member of a wall.

wall tie Steel wire in various shapes designed to either tie brick or stone walls, such as in CAVITY WALL construction and to tie brick veneer to framework. Ties are also made to secure veneer.

wane Bark, or a missing piece of wood from any cause, on the edge or corner of a piece of wood.

washing down Cleaning masonry with a dilute solution of water and muriatic acid or another cleaning material.

waste Short for *waste pipe.* Waste pipes are part of the DWV SYSTEM in plumbing and only carry waste water from lavatories, sinks, tubs, and the like.

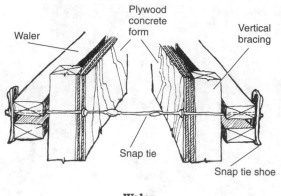

Waler

Waste pipe is made of a number of materials:

- **Cast iron.** This is heavy thick-walled pipe with a variety of forms at the end—possibly straight or bell-shaped (known as the *hub* or *spigot* end). The hubs enable the pipe to be joined by the use of hot lead and oakum (a fibrous material). Cast iron is commonly available in 5' and 10' lengths and in 2", 3", and 4" diameters for residential use. *See also* CAST IRON.
- **Fiber.** This is a gray composition pipe available in various lengths and tapered at the ends. It is commonly used underground for cesspool and septic tank connections, and is connected with rubber gaskets.
- **Plastic.** A black (ABS) or cream-colored (PVC) pipe, depending on the plastic. It comes in 10' and 20' lengths and in 1½" to 4" diameters. It can be used underground but must be carefully supported by solid earth.

water closet   Trade term for a toilet. The name *water closet* comes from the design of early toilets. There was a box, or closet of water high up on a wall. Pulling a chain released the water into a pipe which lead to the bowl, flushing it.

There are three basic kinds of toilets:

- **Siphon jet.** Quiet and powerful, this is the best flushing action available. A siphon jet toilet is characterized by a low, sleek bowl.
- **Reverse trap.** A type of toilet whose trap is at the rear.
- **Water washdown.** Toilet whose trap is at the front. This type of toilet can be recognized by the large bulge in the front of it and is banned in many areas. The reason? There is a relatively small area for the water and a large area in the front that is exposed to contamination. Moreover, it does not flush as well as the other types.

Development of the toilet

For a relatively simple device, development of the toilet took a long time to create, perhaps because it is so closely related to personal bathing. Bathing was—hard as it is to believe—considered anathema for a long time. As a possible overreaction to the Roman baths of antiquity, the Christian church associated unbridled sexuality with bathing. After awhile, bathing was partially banned by the Church, and waste was generally disposed of in the most unhealthy ways possible, including dumping it into castle moats or simply out the window.

The cholera epidemic in London in 1832 focused attention on cleanliness, however, and the toilet finally came into being. In 1178, one Joseph Branch had invented a *syphonoic* or *water closet*, and then in

1891 Thomas Crapper patented an automatic flushing system. The water closets he manufactured were named "Crapper"—a name that ultimately degenerated into common slang. Thereafter, toilets were made of wood lined with copper, zinc, or galvanized steel, but these were hardly sanitary materials.

In 1907 in America, the VITREOUS CHINA toilet was introduced by the Eljer Plumingware Company—a giant step forward in toilet design. When the toilet was introduced, people were pleased with its apparent ability to keep clean, but they wondered about its strength. Was the water closet strong enough to hold almost sixty pounds (6½ gallons) of water? In a dramatic demonstration, Raymond Elmer Crane showed how strong the toilet was by having 27 men with a total weight of 4748 pounds stand on two planks laid on the tank—and it didn't crack. With that illustrious beginning, the china water closet was here to stay.

Water closet

water pipe Pipe use to route POTABLE water through a building.

Many of the same materials used for WASTE PIPE are used for water pipe but simply are available in smaller diameters. For residential use, the average ID is ½"; the main pipe coming into the house is ¾".

The following are materials of which water pipe is made:

- **Galvanized.** This pipe is found in many older buildings. Silvery in color, it is BLACK IRON pipe that has been dipped in zinc—galvanized. Galvanized pipe has one disadvantage: it scales, and this can lead to blockages within the pipe. When replacing a section of galvanized, many plumbers will use copper with dielectric FITTINGS.
- **Copper tubing.** Copper tubing may be straight or coiled, depending on the type. A straight copper comes in 10' and 20' lengths and in diameters ranging from ¼"–2" and in various lengths from 45'–200'. The three types of tubing available are:
    - ~ **Type K.** This comes in rigid lengths and coils and is designed to be buried in the ground.
    - ~ **Type L.** This is used inside the house. It is available in rigid lengths and coils, but the rigid also comes in type LBT, or bending temper. The pipe can be bent with a special bending tool.
    - ~ **Type M.** This is used in heating applications.
- **Plastic pipe.** In recent years, plastic pipe has become much more widely used, and there is a variety to choose from:
    - ~ **CPVC.** This is chlorinated polyvinyl chloride: solvent-welded pipe that can be used for hot- and cold-water systems. Some plumbers say that it doesn't work well with hot water, tending to bend or "spaghetti" if the water is too hot.
    - ~ **PB.** This is tylene, a flexible heat-resistant thermoplastic piping used for hot/cold water supply. It is joined by mechanical fasteners. In addition to being inexpensive, PB supply tubes are "forgiving"—

they can be slightly bent to make a proper connection.
~ **PE.** This is thermoplastic pipe used for cold water, outdoors and below ground.
~ **PP.** This is polypropylene, a semi-rigid thermoplastic with high
chemical and heat resistance used for making tubular drainage
goods. It cannot be solvent-welded and therefore is joined by mechanical fasteners.
~ **PVC.** This is rigid thermoplastic used for DWV SYSTEMS and cold-
water-only pressurized systems outdoors or in the ground. Connections are made with solvent welding
• **Brass pipe.** This comes in various diameters (¾″ is common) and
is available in lengths up to 20′. Brass is good in situations where
water is very corrosive, and it lasts longer than other metal pipe
(not clogging or scaling), so it's a favorite of plumbers in certain
situations—such as the pipe that goes between the water company
pipe and the house water pipe.

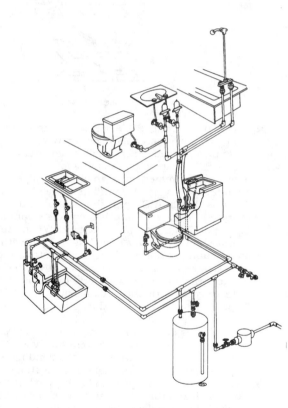

Poly flex connector

water pressure The rate at which water flows in a house. Water comes to a
house either from a city main or a well. It travels under pressure, described
in terms of pounds per square inch, or PSI. Normal pressure varies between 50 and 60 PSI, and should not drop below 30 PSI or go above 80 PSI.

A variety of factors affects the pressure. The size of the supply line, which depends, in turn, on the fixtures the line will serve. Though local CODES may vary, normally, for example, a 1" diameter pipe is required for more than three FIXTURES; a ¾" pipe is acceptable for three fixtures and ½" line is adequate for one or two fixtures.

Each fixture has its own branch line, and this, also, must be properly sized. Normally a ⅜" line is adequate for sinks and lavatories while a ½" line will be used to serve each fixture.

At different times of the year, for example, there is more or less pressure based on overall use. In the summer, for example, with lawn sprinklers on and more water used for pools and drinking, the available pressure of public water will be diminished.

Another is the condition of the water supply lines into the house. If the line is old and a lime buildup caused by hard water exists, the pressure may be reduced.

So, too, pressure will be affected by individual turns within a particular water supply system. The more convoluted the system, the more likely the pressure will be lower. Likewise, the length of the run will also affect pressure: the longer the run, the more likely the pressure is to be lower.

water retentivity The ability of MASONRY to hold water.

water supply system The collection of pipes and valves that combines to deliver POTABLE water in a building.

Water begins its journey to a house from a waterworks (where it is purified) or from a well. Under pressure (See WATER PRESSURE), the water enters a home through a main service line ¾" or 1" in diameter, depending on the size of the system. As it enters, it goes through a meter, which records its usage. Inside, it branches out into a pair of pipes, one of which is routed through the hot-water heating plant where it is heated, and then the pipes travel side by side throughout the house to the various FIXTURES and FITTINGS that use them. Some of these lines travel vertically and are called RISERS; those that travel horizontally to the fixture or fitting are called _branch lines._

At various points in the system are VALVES, which are faucet-like devices meant to shut down water supply to pipes where repair is needed. These are on runs to control specific lengths and are usually (or should be) under sinks and toilets: one each for the hot and cold water, and one for the toilet tank.

Valves are also located at the bottoms of risers, meant to be used to drain the pipes. Additionally, homes also usually have one or two valves at the water meter. One is called a _meter shutoff_ and is on the "street" side of the meter; the other is on the house side and is called the _main shutoff._

Both control water flow to the entire house, but either before or after the water passes through the water meter. _See illustration on the following page._

water supply tube Standard water supply tube, usually called _water supply,_ is for sinks, toilets, and the like, connecting the water supply pipes with these fixtures. It comes in chrome-covered copper, rough copper (no chrome), plastic, and corrugated copper and in small diameters (usually ⅜" or ½"). They are connected to fixtures with compression FITTINGS.

watt A unit of electrical energy.

weather In allowing exterior wood to weather a few weeks or months or more on the assumption that it will make it a better substrate for a finish. _See_ STAINS.

weatherstripping Material used to close off openings around doors and windows.

Weatherstripping comes in a wide variety of forms, from self-adhesive strips secured around doors and windows to coils of caulk. Whichever is used, weatherstripping factors greatly in energy savings and increased

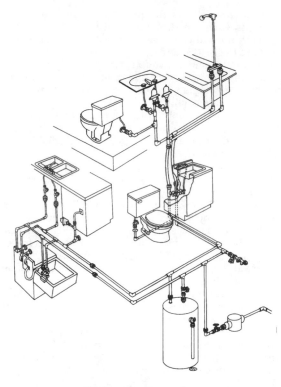

Water supply system

comfort level within a building. Indeed, it is estimated that proper CAULKING and weatherstripping of a house can reduce fuel bills up to one-third, because these things help keep warm air in and cold air out, and heating units don't have to work as hard.

New homes normally come with built-in weatherstripping, but as time passes these things will have to be replaced. *See also* STORM WINDOW.

web 1. The thin center portion of a beam that connects the wider web and bottom flanges. **2.** The cross wall connecting the face shells of a hollow concrete masonry unit.

weep hole In masonry, a hole made in brick so water can run out of MASONRY WALLS rather than collect and do damage. Weep holes are simply vertical gaps in mortar joints in the cavity-type wall and are normally set 24″ apart. To ensure that they're not blocked by excess mortar falling down as bricks are set in place, masons may place a protective board called a *shield* above the weepholes, and then raise it by rope as they work, or bevel the MORTAR BEDS so there is less of a tendency for excess mortar to be squeezed out. *See illustration on the following page.*

welded wire fabric *See* LATH.

well Manmade hole in the ground that produces potable water.

wellhold Space in a building for a staircase, as well as the space around staircase.

western framing Another name for PLATFORM FRAMING.

wet wall PLASTER wall. The name comes from the method of wall installation. A series of wet materials are troweled on, as opposed to DRYWALL, where gypsum boards are used and only JOINT COMPOUND is used.

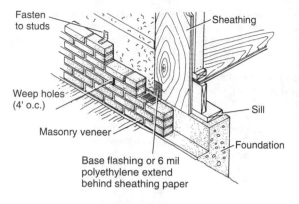

Fasten to studs

Sheathing

Weep holes (4' o.c.)

Sill

Masonry veneer

Foundation

Base flashing or 6 mil polyethylene extend behind sheathing paper

Weep holes

whip Up-and-down movement of UNDERLAYMENT that hasn't been nailed down properly.

white ants Another name for termites—inaccurate, because termites are neither white nor are they ants.

white wire The color of the neutral wire in electrical installations.

wicket A small door within another door.

wicking The tendency of wood to draw moisture through its cells by capillary action in the direction of the grain (which explains why wood is not placed in contact with the ground; water would be drawn in, and provide a ripe habitat for DECAY to occur).

widow's walk A small deck that straddles a roof. In whaling days, when seamen went off for three- and four-year stretches, their wives used to climb up in the widow's walk to see if the ships were returning. Since many fishermen didn't return, the nickname evolved.

wind break *See* FLASHING.

wind load The pressure that the wind exerts on a structure. This is another facet (*see* DEAD LOAD, EARTHQUAKE LOAD, *and* LIVE LOAD) in calculating just how strongly a house should be built. Ordinarily, engineers specify that a house should be built to take an average pressure of about 15 pounds per square foot, which would be the equivalent of a 75–80 mph wind.

Factored into this, of course, is the shape of the house. Some houses, such as a Cape Cod, will not take the wind as well as one with angular shape where the wind does not impact with direct force.

It is possible to design a house, bolstering it with iron STRAPPING, for example, and FRAMING FASTENERS, to stand up to hurricane-force winds. Designing against a tornado, though, is not practical. It is estimated that the winds inside a tornado may be swirling as much as 500 miles an hour, which translates into about 75 pounds of pressure per square foot.

Taking off the roof

It is assumed that excessive wind force blows shingles off the roof. In fact, it is what engineers call "negative pressure" that does it. As the wind hits the house broadside, it sort of slides off and wraps around the roof, gaining velocity as it does, and as it does the pressure on the house fall, creating a sucking or negative pressure that pulls the shingles off.

winder A step cut into a long, triangular shape. Winders are used in winding staircases. The TREAD shape enables them to be installed so the stairs can turn as one ascends or descends.

winding stair Another name for a WINDER.

window Unit with panes of glass encased in a framework of some sort. Windows, commonly called lights in the trade, are composed of basic framework of parts consisting of a SASH or the movable parts in a framework that houses them. The sills and other trim pieces are added after the window is installed.

Windows are available in a wide variety of types, made of different materials, and in either stock or replacement types. Stock windows come in sizes made to fit into an existing ROUGH OPENING, and in remodeling work normally don't fit exactly. For example, an existing opening may be 24″ wide and the closest stock size might be 26″. (In this case, the contractor would have to install filler pieces between the window and rough opening.) Actual replacement windows are made to fit exactly.

- **Wood.** The classic wood window is pure wood inside and out and ranges in quality from a "builder's quality" to good quality.
  (The heavier the wood members, the higher the quality.) They are available in a wide range of sizes that graduate in 2″ increments. A few companies also make replacement windows.

  Installation varies in complexity, depending on siding and the size of the window being installed. To clean out the old window, however, some siding is removed because it will overlap the edges of the window. The standard procedure is to remove the siding carefully and then replace it once the new window's in.

  Aluminum siding is particularly problematic. It has to be cut extensively to take the old window out, unless a smaller window is being installed, in which case a new framework can be built for it.

  Other types of siding, such as CLAPBOARD and SHAKES, do not represent much of a problem, nor does asbestos (though, to be sure, the siding will be damaged during removal).

  MASONRY also presents a problem because it is a solid material that has to be cut.
- **Wood clad with vinyl or aluminum.** This window is made of wood, with a thin jacket of wood or vinyl. Vinyl-clad windows usually only come in brown or white (though at least one manufacturer makes them in various colors), but aluminum-clad units come in a wide variety of colors. Such windows never have to be painted outside. Most manufacturers offer clad windows in stock sizes only, though custom builts are available.
- **Solid aluminum.** Solid aluminum windows are also available in a range of stock sizes, but they are viewed with skepticism because they conduct cold, which leads to CONDENSATION. This leads to water, and the result is pitting, corrosion, and other maladies to which metal is prone to when water runs over it. Some aluminum windows have thermal breaks in them to prevent condensation, but the consensus among contractors is that they aren't very effective.
- **Pure vinyl.** Vinyl windows are commonly used as replacement windows, and for the simple reason that they are easy for a manufacturer to construct (and thus profitable). The vinyl—PVC, or polyvinyl chloride—can be easily extruded into the components necessary to make the windows, unlike wood or metal.

  Vinyl windows vary greatly in quality. Some fly-by-night companies buy the components and assemble them rapidly, and it's not long before they start falling apart. Other companies make good vinyl windows, typically ones with welded mitered corners rather than with mechanical fasteners. Good vinyl windows are also thicker, with parts 2.33 millimeters compared with 1.5 millimeters found in lesser windows.

  Vinyl replacement windows are usually cheaper and easier to install than other types because the framework stays in place. The sashes and stop MOLDING are removed and the new window, made to the exact size of the opening, is slipped into place, secured, and CAULKED (or a window almost the exact size needed is slipped into place and fillers are used around it).

Of necessity, the new window will be smaller than the existing, simply because the first window's framework is staying in place—it's as if you remove a picture from its frame, then install a new picture with a new frame inside it. Essentially, a new window is being installed inside an existing one. Replacing an aluminum window with a vinyl one is more problematic. Metal sections are difficult to remove—the entire window has to come out.

Many contractors, despite cost savings and installation ease, however, don't use vinyl windows due to the expansion and contraction of the vinyl, which breaks the caulk seal around the window and allows for drafts and leakage.

window frame General term used to describe all of the parts of a window except the glass.

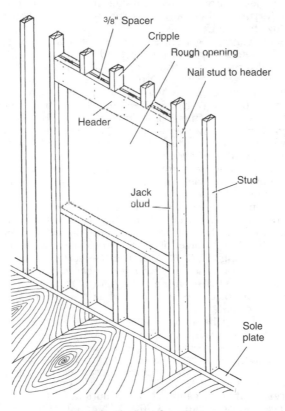

Rough window framing

window sill The flat shelf-like exterior part of a window. The sill is often confused with the STOOL, which is actually the corresponding interior part of a window.

window wall Wall in which much of the area is composed of windows.

windshake In wood, separation of the annual growth rings in a tree. This defect in wood is thought to be caused by the tree being pulled from the ground by wind—hence the name. Since it is cup-like in appearance, it is often called CUPSHAKE.

wiped joint A type of joint used in connecting cast iron piping. A wiped joint is made with hot lead and not without some skill. (Some plumbers specialize in teaching this.) These days, wiped joints are not much used anymore, but plumbing codes still require plumbers to be able to do it.

wire glass Glass with thin wires embossed in it. Wire glass is commonly used in factories and the like, both as protection against intruders and protection from broken glass; a shattered wire glass window won't shatter, as the mesh holds the glass in place.

wire lath A wire used in CONCRETE reinforcement.

wire mesh A webbed steel used in CONCRETE reinforcement.

wire nut A device used to connect wires. Wires nuts, also known as SOLDERLESS CONNECTORS, and splice connectors are commonly used in making connections in house wiring. They consist of a plastic shell inside of which is insulation and a coil of wire. They are simple to use: the bare end of the wires to be joined are inserted, parallel, into the bottom of the wire nut, which is then turned clockwise. As the nut is twisted, the two wires are twisted together securely.

wire ties Short lengths of galvanized or cement-coated wires used for tying masonry, wood, and metal members together.

wiring General term describing the electrical wiring in a building. Wire for electrical purposes is called CONDUCTORS, simply because it conducts electricity. (Actually, wire is a misgnomer, because you can't buy bare metal. Rather the wire—the metal—comes covered with insulation, usually plastic, and is also commonly known as CABLE.) The conductor may consist of solid metal or twisted strands. Solid conductors are used in most house wiring, while the stranded conductors are used in CONDUIT wiring and other applications where flexibility is important.

Conductors these days are always copper; aluminum was once used, but its common expansion and contraction when heated led to loose connections, sparking, and fires.

Wire is commonly described in terms of number of the wire: the larger the number the smaller the wire. No. 38 wire, for example, is a little thicker than a human hair; No. 18 wire is about the diameter of the head of a pin; No. 2 wire has the diameter of a lead pencil. The designation can become even larger, moving to 1/0, 2/0, 3/0, and so forth.

The bigger the wire, the more current it can carry (*see* AMPACITY). Those commonly used in house wiring are numbers 14 and 12. Numbers 18 and 16 are common for extension cords, while number 14 is used for heavy-duty appliance cords.

Insulated wire (what other type is there?) is also classified by letter according to the type of covering or insulation used. *See also* BX, ROMEX.

Following is a list of common types of wire:

- **Low voltage.** This is a thin wire of 18, 20, or 22 gauge. It is designed for use with devices that require low voltage, such as bells, outdoor lights and THERMOSTATS.
- **Service entrance cable.** This is neoprene-coated black wire and comes in various gauges: 0, 00, 1, 2, and 3. This cable is used between the ELECTRIC SERVICE head and the house.
- **THHN.** This is nylon-covered wire, available in various sizes and colors and both oil- and heatproof. It is a slippery, abrasion resistant material and can easily be pulled through CONDUIT. THHN is wire used for fluorescent fixtures because the ambient temperature is 105 degrees, which is higher than NM or BX cable could readily stand.

• **UF.** The letters stand for underground feed cable. This is a strong abrasions-resistant material that is waterproof and can withstand damage. As such, it is used underground.

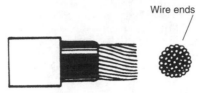

Wire ends

2/0 Copper wire

Actual size of No. 2 copper wire

wiring, surface Electrical wiring mounted in flat metal sleeves which, in turn, are mounted on walls. Surface wiring, often called *raceway wiring*, is a simple solution to adding outlets when going through walls would be difficult.

woodwork General term that describes the TRIM on the inside or outside a house.

The guys with the gray hair

As carpentry jobs go, the easiest jobs are those involving the erection of a framework for something and adding woodwork is the most difficult. Wood pieces that comprise it often must be cut at angles and for appearance must fit precisely. Despite the availability of stationary power tools to ease the job, it still takes experience. As one carpenter put it, "The young bucks will be doing the framing on the house, but it's the guys with the gray hair who do the trim."

working to one's limit The way a mason works: a right-handed mason will usually lay units working from the left corner to the center and the reverse is true for a left-hander.

wreath A curved section of a staircase used to join the NEWEL post with the rising run or handrail.

wye A plumbing FITTING shaped like the letter Y.

wythe *See* BRICK.

yard A cubic yard of concrete. This is the basic measure of concrete.
yellow hat A wire connector. These devices are different colors according to the size wires they can handle.

z bar Wire from 4″ to 6″ long, shaped in the letter Z and used to tie the interior and exterior walls of a CAVITY WALL together

z flashing Z-shaped piece of galvanized steel, aluminum, or plastic installed at horizontal joints of plywood siding to prevent water penetration.

zoning laws Rules and regulations established by local government concerning where buildings may legally be constructed. Cities and towns regulate who can build where, and what, and what PERMITS are required. To build without knowing the zoning laws is to risk everything from the project being stopped by a town inspector to having it torn down after completion. *See also* VARIANCE.